Multiple Intelligences and Standards-Based Mathematics

Hope Martin

Arlington Heights, Illinois

Multiple Intelligences and Standards-Based Mathematics

Published by SkyLight Professional Development
2626 S. Clearbrook Dr., Arlington Heights, IL 60005-5310
800-348-4474 or 847-290-6600
Fax 847-290-6609
info@skylightedu.com
http://www.skylightedu.com

Senior Vice President, Product Development: Robin Fogarty
Director, Product Development: Ela Aktay
Senior Acquisitions Editor: Jean Ward
Editor: Barb Lightner
Cover Designer and Illustrator: David Stockman
Book Designer: Bruce Leckie
Production Supervisor: Bob Crump
Proofreader: Ann Wilson
Indexer: Candice Cummins Sunseri

Printed in the United States of America.

LCCCN 99-75654
ISBN 1-57517-185-6

2547-McN
Item Number 1882

Z Y X W V U T S R Q P O N M L K J I H G F E D C B
06 05 04 03 02 01 15 14 13 12 11 10 9 8 7 6 5 4 3 2

To my family—for their love and support.

Acknowledgments

As a teacher, I have met some very interesting people, both students and colleagues. We've shared philosophies of education, discussed thoughts for new lessons, and ruminated over novel strategies to teach them. How exciting and thought-provoking these discussions have been!

But on occasion, I have met people who have done more than share their ideas; they've helped to change the direction of my life. I would like to thank two of these people.

The first is an extraordinary teacher, Dr. José DeVincenzo. I remember thinking when I first entered his classroom, "Another class in cognitive psychology—how exciting!" How wrong I was! José introduced me to the theories of Howard Gardner and made his ideas real. As a mathematics teacher I was well aware of the limited success mathematics teachers have had using traditional programs. I constantly was changing my techniques and strategies in an attempt to better serve my students. I examined standards-based mathematics, constructivism, cooperative learning, and, now, I was examining multiple intelligences. Later, I came to realize that our discussions of Howard Gardner's *Frames of Mind* (1983) would serve as the foundation to bring focus and direction to my mathematics program.

The second person who significantly changed my "way of doing things" was Jim Bellanca. Jim had seen and liked some materials I developed for a workshop. He thought we should talk. And so we met to discuss how to take my filing cabinets full of "math stuff" and organize them into a book. Jim asked me to describe some of my activities and strategies and, as we talked, I realized I did have a personal philosophy of education. What I taught and how I taught was not as capricious and disorganized as I believed it to be—it was firmly rooted in the NCTM standards and the cognitive theories of Howard Gardner. The changes that started in a classroom in Evanston, Illinois, were finalized in a meeting room in Palatine, Illinois. Jim not only gave me the opportunity to have my ideas published, he gave me direction by sharing his experience and knowledge.

I wish to thank both José and Jim for sharing their knowledge and experiences and helping me do what I love most—teach through my creations.

I also want to thank Arnie Martin and Joel Cohen, two talented musicians who helped me over some difficult roadblocks while writing this book. They increased my understanding of music theory and helped give substance to the musical activities.

Contents

Introduction

Mathematics is the key to opportunity. . . . For students it opens the doors to careers. For citizens, it enables informed decisions. For nations, it provides knowledge to compete in a technological economy. (National Research Council 1989, 1)

STANDARDS-BASED MATHEMATICS: THE FOUNDATION OF AN EXCELLENT CURRICULUM

In 1989, the National Council of Teachers of Mathematics (NCTM) published the *Curriculum and Evaluation Standards for School Mathematics* in which they encouraged educators to re-examine the mathematics curriculum and pedagogy and to make necessary changes so that mathematics can become more than a "spectator sport." In the past decade, policy-makers have worked to democratize the mathematics classroom, make advanced mathematics accessible to more students, and encourage all students to study more math. In response, school districts across the country have analyzed their math goals and objectives and developed documents that align their curriculum to the national standards outlined by the NCTM. They have expanded the mathematics content of their schools to include topics that generally appear "at the back of the book." They have studied ways to supplement and enrich their students' mathematical experiences. There has been a proliferation of "replacement units" offering hands-on, active approaches. In addition, schools are examining authentic assessment in order to develop more comprehensive evaluation programs.

However, "certain messages of the original *Standards* documents were not understood" (NCTM 1998, 13) and in response to these concerns, the NCTM has drafted *Principles and Standards for School Mathematics (Standards 2000)*. This document builds on the principles submitted in the original documents and is being designed to lead us into the new millennium. The initial draft of *Standards 2000* emphasizes the interplay between mathematical processes and the learning of mathematical content: "With an emphasis on understanding mathematics that is fundamental, educators can be supported as they move beyond some of the superficial interpretations that have been made of ideas from the previous *Standards* documents" (NCTM 1998, 17).

Some of the guiding principles of the draft of *Standards 2000* are

- Developing worthwhile mathematical tasks.
- Making mathematics available to all students (equity of mathematical opportunity).
- Designing programs that are responsive to the intellectual strengths and personal interests of students.
- Planning assessment programs that monitor student progress and can be used to inform teaching.

The document asks, "What mathematical content and processes should students know and be able to use as they progress through school?" (NCTM 1998, 45). It proposes ten standards: Five of the standards are content standards (what mathematics should all children learn?). These address the areas of number, measurement, algebra, geometry, and data. Five are process standards (how should students acquire and use mathematics?). These address problem solving, reasoning, connections, communication, and representation.

The topics in *Multiple Intelligences and Standard-Based Mathematics* have been aligned with the proposed content standards, and each chapter reflects the topics, as outlined by these standards. In addition, each of the lessons incorporates the process standards and requires students to problem solve, reason mathematically, communicate, make connections, and use representations to organize, record, and model their solutions.

In the working draft of *Standards 2000,* the NCTM defines a technology principle in the following way: "Mathematics instructional programs should use technology to help all students understand mathematics and should prepare them to use mathematics in an increasingly technological world" (NCTM 1989, 40). In the years following *Standards,* the explosion of technological advances (graphing calculators, personal computers, and access to the World Wide Web) has encouraged schools to incorporate technology into their mathematics programs. The activities in *Multiple Intelligences and Standard-Based Mathematics* include a section, "Tie to Technology," that encourages students to utilize calculators, computers, geometry programs, and/or web browsers to supplement their learning. As stated by the NCTM, "Technology can make mathematics and its applications accessible in ways that were heretofore impossible" (1989, 40).

The NCTM has expressed concerns about the equity of mathematics programs across the country. While verbal literacy is expected of all students in the United States, all too often, educators are willing to believe that only *some* children can learn mathematics. It no longer is an option to teach mathematics to a limited number of American students. However, many schools have addressed the equity issue—teaching mathematics to all students—by placing students into differentiated groups, lowering their expectations, and revising curriculums. As Lawson, a retired member of the California State Board of Education, states:

> Opportunities to learn are not committed or distributed fairly among United States students. By the seventh grade, classes and topics have become quite differentiated, as students who are perceived to be weak in mathematics are moved to remedial groups or given less challenging subject matter. Some students are forever relegated to a slower pace or to different courses, depriving them of learning opportunities accorded to other students (1990, 2).

"Adopting the belief that all students can learn mathematics is critical" (NCTM 1998, 24). How do we achieve equity in the mathematics classroom? The NCTM suggests that an excellent mathematics program would provide "solid support for . . . learning" (1989, 26), and among other things, be responsive to "intellectual strengths" (1989, 26). When mathematics acts as a gatekeeper and our system sorts students into two groups—those capable of learning and those incapable of learning—we limit the future educational and employment opportunities of students.

To design programs that are responsive to the intellectual strengths and personal interests of students, we must fully explore the alternatives to traditional mathematics instruction. We need to examine not only what is taught but also how it is taught. To do this we need to explore cognitive theories and how children learn.

THE ROLE OF MULTIPLE INTELLIGENCES IN TODAY'S MATHEMATICS CLASSROOM

To make mathematics available to all students and to be responsive to the intellectual strengths of students, we need to examine the advances that have been made in the field of cognitive psychology. Howard Gardner theorized the existence of multiple intelligences to explain our intellectual potential. In his seminal book, *Frames of Mind* (1983), he encourages people to expand their definitions of intelligence, scrutinize their attitudes toward learning, and expand their beliefs about the way children learn. Relying on a framework of cognitive and developmental psychology, Gardner presents a case for the existence of multiple intelligences.

A careful examination of Gardner's theory of multiple intelligences by mathematics educators would most certainly influence the pedagogy and/or mathematical *process*. By utilizing Gardner's theories and incorporating them into the mathematics classroom, educators can support the strengths and shore up the weaknesses of their students.

How do multiple intelligences play out in the mathematics classroom? Following is a brief explanation of each of the intelligences identified by Gardner (1983, 1995).

Verbal/Linguistic Intelligence

This is the intelligence of language and words. Students who are strongly linguistic love to read, write, and discuss ideas. Educators can encourage students to develop their verbal/linguistic intelligence by requiring both symbolic and verbal solutions to problems. All of the activities in *Multiple Intelligences and Standard-Based Mathematics* provide a journal question to encourage this intelligence.

Logical/Mathematical Intelligence

This is the intelligence of numbers and logic. Students whose strength is this intelligence love to problem solve and use symbolic abstractions. Activities that relate to real-world mathematics encourage students to develop this intelligence by demonstrating the power and usefulness of mathematics.

Visual/Spatial Intelligence

People whose strength is visual/spatial understand their physical world. Research has indicated that there appears to be a correlation between the mathematical performance of females and their spatial intelligence: the higher a girl's spatial skill, the higher her mathematical achievement (Tartre 1990). To help students see things in images and pictures, teachers can encourage the use of visual mathematical representations. These are provided in many of the activities contained in this book.

Musical/Rhythmic Intelligence

This is the intelligence of rhythms and melodies. While each of us holds musical capabilities to some degree, in some societies, students are encouraged to develop their musical intelligence to a heightened degree. For example, in Japan, musical/rhythmic intelligence is considered a vital part of the educational process, and thus most students study Suzuki, a method of learning to play the violin. While American educators do not pursue the development of this intelligence to the same degree, we can encourage our students by taking advantage of the connection between rhythms and fractions and putting important mathematical concepts and rules to music. The Collection of Math Medleys address this intelligence in a unique way.

Bodily/Kinesthetic Intelligence

People with a strong bodily/kinesthetic intelligence are hands-on learners who enjoy tactile experiences. By using manipulatives and encouraging the concrete, physical representation of mathematics concepts, teachers can improve bodily/kinesthetic intelligence and bring comfort to those students who excel in this area.

Interpersonal Intelligence

Those who possess an enhanced interpersonal intelligence love working with other people. The activities and projects presented in this book are designed to foster collaboration and they encourage students to work together to solve a mathematical problem. These lessons cultivate the growth of interpersonal intelligence as they develop deeper math concepts.

Intrapersonal Intelligence

This is the intelligence of the inner self. Understanding and being in touch with one's feelings and thoughts are at the center of this intelligence. By encouraging students to explain their reasoning and thinking, they become more self-reflective and develop their intrapersonal intelligence.

Naturalist Intelligence

In the mid-1980s, Gardner suggested an eighth intelligence—the naturalist. He describes individuals who possess this intelligence as people who can classify, order, and define objects based upon common attributes. Children who can "make acute discriminations among cars, sneakers, or hairstyles" (Gardner 1995, 206) exemplify naturalists in our modern society. This intelligence can be encouraged in the mathematics classroom by using manipulatives that challenge students to find common attributes, Venn diagrams, and logic activities to classify objects by their characteristics.

THE ROLE OF ASSESSMENT IN THE NEW MILLENNIUM MATHEMATICS CLASSROOM

Students who are participating in worthwhile mathematical tasks are active and are discussing, questioning, modeling, reasoning, and problem solving. Traditional evaluation strategies would reveal to a limited degree the learning that is taking place. Educators are experimenting with alternatives to traditional assessment as a complement to typical testing. What is the difference between assessment and testing?

The *Encyclopedia of Educational Evaluation* defines assessment in the following way:

> Assessment, as opposed to simple one-dimensional measurement, is frequently described as multitrait-multimethod . . . it focuses upon a number of variables judged to be important and utilizes a number of techniques to assay them (Anderson et al. 1975, 27).

With the development of state and national mathematics standards, traditional testing has come to define mathematics assessment in many parts of the country. These tests generally measure specific, discrete skills and students are asked to recall certain facts. While these tests are being used to evaluate students' progress, they generally are not used to influence the instructional process. For this reason, it would be futile to embark on a campaign to change the quality of mathematics instruction without exploring alternatives to these more traditional evaluation instruments. Traditional testing and more nontraditional assessment techniques should be looked at as partners rather than adversaries. Both can be used to help maintain high standards, to follow the academic progress of students, and to help teachers modify old and develop new curriculums to better meet the learning needs of today's students.

The view of mathematics presented in *Standards* (NCTM 1989) defines goals for evaluation as process goals that students work toward on a continuous basis. Process goals cannot be measured accurately using any one assessment technique since students should be viewed over a period of time to best appraise their mathematical growth. Sometimes referred to as authentic assessment, performance tasks can be aligned with the goals and objectives of the mathematics curriculum and are more open-ended rather than being tightly structured problems. The tasks provide rich opportunities for students to display their reasoning and thought processes. Following are some nontraditional assessment techniques that can be very useful in the process of evaluation and assessment.

Observation and Questioning

Observation affords the teacher the opportunity to diagnose understanding, evaluate a student's participation in small group activities, and assess the problem-solving strategies being used and the mathematical communication skills of the group members. Questioning, such as "How would you explain this problem in your own words?" or "Why do you believe this solution is correct?," develops reasoning skills and encourages students to understand the meaning of "mathematical power." When student observation is suggested as part of the assessment of an activity, sample questions usually are indicated.

Journals

Written communication is an important part of connecting mathematical skills and concepts. All too often, teachers are overly concerned with the symbols and procedures of math.

Using journals provides teachers and students the opportunity to make connections between explaining a mathematical concept and understanding the concept by using symbols, models (visual and manipulative), and language. A model of this relationship might look like this:

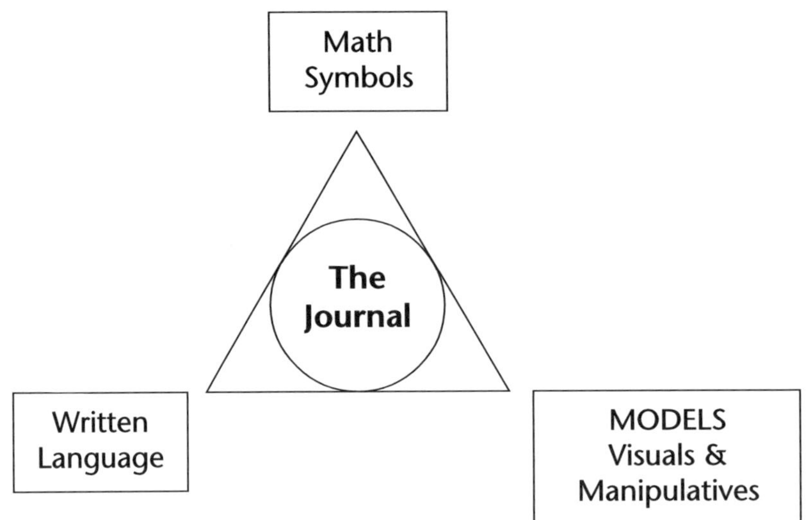

Journals generally are used to measure affective issues, but they can also be very helpful in defining the depth of a student's understanding. When students are asked to solve a problem using mathematical procedures and then draw a picture or explain in their own words how they solved it, they are using a variety of intelligences and improving their understanding. Sample journal questions are part of each of the experiences presented in this book.

The Grading Matrix

The grading matrix is a unique means of correlating the goals and objectives of the lesson with evaluation. Rich experiences require students to perform many different tasks. In the same lesson, students may be asked to measure, perform computations, work collaboratively to problem solve, and present their findings on a graph or chart. Each of these tasks can be evaluated, and distinct evaluations can be made.

Using these techniques affords teachers the opportunity to understand how well the students understand the mathematics, to respond to the students' thinking, and to revise or modify instruction if necessary to better meet the needs of their students.

USING THE TEACHER'S PAGE

The mathematical experiences in this book are aligned to the NCTM's process standards—problem solving, communication, connections, reasoning, and representation—and content standards—number and computation, algebra, measurement and geometry, and data analysis, statistics, and probability. Each activity is preceded by a Teacher's Page that provides valuable information for managing the lesson. Teachers also can use these pages to record any changes or note connections to other activities and/or lessons for future use.

Following are brief descriptions of each of the sections of the Teacher's Page:

Math Topics: As in most hands-on activities, the mathematical experiences presented in this book address more than one math skill or topic. In the real world, mathematics is an integrated experience with skills and concepts that interrelate and blend. When planning the activities, teachers can use this section to connect the lesson to skills and concepts that are part of their mathematics curriculum.

Types of Intelligences: Because of the open-ended nature of many of these activities, students are encouraged to explore their multiple intelligences and reinforce some while strengthening others. These rich activities usually address more than one type of intelligence.

Concepts: This section gives an overview of what students do and describes the processes involved in the activity.

Materials: This section provides a comprehensive list of materials and supplies, which should be gathered and made ready for the lesson. This will help the activity run more smoothly.

What to Do: This section is not intended to be a scripted, step-by-step plan but rather suggestions to help encourage students and act as motivation for the experiences. Some lessons do suggest specific questions or how to develop student understanding. These are merely suggestions and should be used only if appropriate to the needs of the class. Some problems are open-ended in nature and have more than one possible answer. Thus, no answers are provided for them. With less open-ended problems, answers are provided in this section.

Variation: Some activities can be extended or elaborated upon and suggestions are offered in this section. If the teacher has additional activities that can be added to enhance the lesson, this would be a fine place to write them down for future use.

Tie to Technology: Using graphing calculators and/or spreadsheets will help students move into the mathematics needed for the future. Calculator use is suggested when calculations are difficult or when graphs would better explain the lesson. Computers are helpful when using formulas that require repetitive calculations.

Graphing Calculators
There are many graphing calculators available and while each one has unique functions they do have common advantages. In addition to performing the functions of most scientific calculators, many graphing calculators can be programmed with formulas and complex procedures. Calculators with table capabilities can be used to perform complex statistical functions.

Computers
The invention and use of computers has, more than any other discovery, led to the explosion and availability of information. Spreadsheets can be used in the mathematics classroom to make tedious and repetitive calculations child's play. Students learn the mathematical formulas as they insert them into the program. Regardless of the complexity of the spreadsheet program, they all are capable of doing intricate calculations quickly and efficiently. They can also, in the blink of an eye, produce diverse graphs of the data. Students can experiment with different graphs to find the most appropriate one. When the use of a spreadsheet is suggested, two examples are supplied on the Teacher's Page. One is a sample of a "finished" spreadsheet and the other contains the formulas that might be entered into each cell. These are merely suggestions of possible formulas.

Computer Software
Another recent computer innovation is the development of the PowerPoint program. When used in conjunction with Internet websites, it allows students to research a topic, import pictures (from the website) directly into their presentation, and prepare a professional looking presenta-

tion of their research. For example, students can research (via the Internet) the accomplishments of great mathematicians throughout history. They can import pictures and diagrams into their presentations.

Assessment: Multiple suggestions are made here. They may include traditional quizzes or tests, completed student products, observation and questioning, suggested journal question(s), and the Grading Matrix.

On the Internet: If an interesting website is available, it is described in this area. (The site's URL is listed in the appendix.) Since information and sites on the Internet frequently change (some on a daily basis) and frequently are expanded or eliminated, the URLs should be checked and revised when necessary. This type of cyberspace research allows teachers to develop interdisciplinary units.

It is essential that these experiences fit into the curricular goals of the school and the teaching style of the people using them. By always remembering to keep the learning needs of students as the primary focus, additions to activities or other modifications can be recorded on the Teacher's Pages, thereby, making the lesson unique and individual!

USING THE STUDENT ACTIVITY PAGES

Student activity pages are provided for each activity. If students are working in groups, generally one copy should be made for each group. Information in the Materials section indicates how these pages should be used. When the activity calls for the collection of group data, it is suggested that an overhead transparency be prepared and all of the data collected be entered on this sheet for further analysis. If an overhead transparency is made of student pages, these can be used to give directions. This is not always necessary and can be used at the teacher's discretion.

USING THE GRADING MATRIX

The grading matrices provided for each activity are based upon the premise that a rubric has been established that clearly defines and distinguishes between levels of performance. The items contained in each matrix evolve from the lesson's objectives and thus more effectively assess student learning. The rubrics are analytical since they clearly distinguish major skills, concepts, and emphases of instruction. Each matrix is task specific. The ratings are based upon the following categories of assessment:

4. Superlative Response: Work is complete in every way with clear and insightful explanations. Mathematics is correct and student (or group) has examined and satisfied all of the conditions of the problem.

3. Competent Response: Work is reasonably complete; explanations are clear. Mathematics is correct and student has satisfied all of the conditions of the problem.

2. Limited Response: Work is partially complete but explanation is not clear and lacks detail. Mathematics has minor flaws and only some of the conditions of the problem have been satisfied.

1. Inadequate Response: Work is incomplete and explanations are not understandable. Mathematics has major flaws and very few of the conditions of the problem have been satisfied.

When developing the matrix, one can assess the following:

1. Accuracy of mathematics.

2. Completeness of work.

3. Quality of mathematical reasoning.

4. Overall quality of work.

5. Affective issues of student collaboration.

Math Skills by Activity	Algebra Concept	Calculator	Computation	Conversions	Critical Thinking	Data Collection Statistics	Estimation/Rounding	Geometry	Measurement	Number Systems & Theory	Probability	Problem Solving	Communication	Math Connections	Reasoning	Spreadsheet/Computer
The Bowling Game	•	•	•		•					•		•	•		•	
All About the Moon		•	•	•	•		•	•		•		•	•	•	•	
Cooking and Mathematics		•	•	•			•		•	•		•	•	•		•
Target Game		•	•		•		•			•		•	•		•	
The Pattern Tells It All					•					•		•	•	•		•
Eratosthenes & the 500 Chart			•		•					•		•	•		•	•
Math and Music			•		•					•		•	•	•		•
Diagramming Divisibility					•					•			•		•	•
The Irrational Spiral	•							•					•	•		•
Vet Math	•		•	•		•						•	•	•		•
Seeing to the Horizon	•		•		•	•	•			•		•	•	•		•
Falling Objects	•			•	•		•			•		•	•	•	•	•
The 18-Hour Clock	•		•							•		•	•		•	
The Bouncing Ball	•	•				•			•			•	•	•		
Soda Pop Math					•			•	•			•	•	•	•	•
The Dartboard	•	•	•		•			•	•		•	•	•	•		•
Math and Music: Frequencies			•				•		•	•		•	•	•		•
Paper-Folding Polygons								•					•	•		•
The Valley of Mars			•	•	•	•	•		•				•	•		•
Limitless Lung Power	•		•			•	•	•	•	•		•	•	•	•	•
Tessellations								•				•	•	•		•
Predicting Colors		•	•		•	•			•	•		•	•	•	•	•
Phone Home		•	•		•	•			•			•	•	•		•
How Long Is Your First Name?	•	•	•			•				•		•	•	•	•	•
Shoe Length vs. Height	•	•				•			•	•		•	•	•		•
Millions of Marshmallows		•	•			•	•			•			•			
The Case . . . Telephone Numbers			•			•					•		•	•	•	•

Multiple Intelligences by Activity	Logical/ Mathematical	Verbal/ Linguistic	Bodily/ Kinesthetic	Naturalist	Visual/ Spatial	Interpersonal	Intrapersonal	Musical/ Rhythmic
The Bowling Game	•	•				•	•	•
All About the Moon	•	•				•		
Cooking and Mathematics	•	•	•			•		
Target Game	•					•	•	
The Pattern Tells It All	•	•		•	•	•		
Eratosthenes & the 500 Chart	•	•			•			•
Math and Music	•	•			•			•
Diagramming Divisibility	•			•		•	•	
The Irrational Spiral	•		•		•	•		•
Vet Math	•	•			•	•	•	
Seeing to the Horizon	•	•	•			•		
Falling Objects	•	•	•			•		
The 18-Hour Clock	•		•		•		•	
The Bouncing Ball	•	•	•		•	•		
Soda Pop Math	•	•	•		•	•	•	
The Dartboard	•	•	•		•	•		
Math and Music: Frequencies	•	•			•	•		•
Paper-Folding Polygons	•	•	•	•	•			
The Valley of Mars	•	•	•		•	•		
Limitless Lung Power	•	•	•			•	•	
Tessellations	•	•	•		•			
Predicting Colors	•	•	•			•		
Phone Home	•	•	•			•	•	
How Long Is Your First Name?	•	•	•			•	•	
Shoe Length vs. Height	•	•	•			•	•	
Millions of Marshmallows	•	•				•	•	
The Case . . . Telephone Numbers	•	•				•	•	

CHAPTER 1

Number Theory, Numeration, and Computation

CHAPTER 1

Number Theory, Numeration, and Computation

There is nothing so troublesome to mathematical practice . . . than multiplications, divisions, square and cubical extractions of great numbers . . . I began therefore to consider . . . how I might remove those hindrances.

—John Napier

Number theory, numeration, and computation are important components of the current school mathematics curriculum. While computation and basic number facts have been emphasized to the detriment of other strands of mathematics, student proficiency in these areas is essential. With guidance and meaningful experiences, students will gain a sense of number, improve their abilities to problem solve, and develop useful strategies to estimate reasonableness of answer. By providing opportunities to examine computation, estimation, and number theory, the activities and projects in this chapter encourage the development of number and operation sense in students.

Understanding order of operations is an essential component of school mathematics. The Bowling Game gives students the opportunity to practice and perfect their skills in an interesting way. The roll of a die makes the availability of numbers the result of chance, and individual creativity and imagination are rewarded in points gained. The lesson can be augmented by using a little ditty, "O^3—Order of Operations Song," to make learning mathematical rules even more fun!

All About the Moon demonstrates the connections between math and science while presenting computation, estimation, conversions, and open-ended problem solving as inviting activities. Interesting moon facts are explained in a way that makes sense to students, and very large numbers become more understandable. In addition, this activity allows students to work collaboratively, which helps them develop their interpersonal and intrapersonal intelligences.

Showing how school mathematics relates to real-world mathematics is a marvelous way to motivate students and keep their interest. Cooking and Mathematics: Converting Recipes shows how understanding fractions is essential to knowing how to cook for a crowd. Why not bake the cookies and make the lesson more interesting? A delicious lesson!

All too often, students operate on numbers using memorized rules and have little understanding of the relative size of the numbers or the reasonableness of the answers they obtain. By using estimation to answer questions such as "Does this answer make sense?" or "What happens when we multiply a number by a decimal that is less than 1?," students develop number sense and become more powerful mathematicians. The Target Game gives students the opportunity to work collaboratively to practice multiplication of decimals and estimate their products.

The patterns found in numbers can be translated into fascinating visual patterns showing the relationship between the logical/mathematical and visual/spatial intelligences. In The Pattern Tells It All, students use number grids to show how prime and composite numbers are visually different. And by writing about their patterns, students extend the lesson to expand their verbal/linguistic intelligence.

Eratosthenes and the 500 Chart relates prime and composite numbers to the rules of divisibility. What appears to be an arduous task is simplified by taking advantage of the power of the patterns of mathematics. To encourage logical/mathematical, visual/spatial, and musical/rhythmic intelligences include the song "The Divisibility Ditty" in the lesson.

Math and Music makes meaningful connections between important mathematics concepts and musical/rhythmic intelligence. Students learn the symbols of music while practicing rational number concepts in a meaningful way. For students with a strong musical/rhythmic intelligence, this activity is especially appealing.

Diagramming Divisibility employs a visual model to encourage students to learn divisibility rules. Some of the number combinations are divisible by two numbers, some by two or three numbers, some by all three numbers. As students problem solve where to place the numbers, they are strengthening their logical/mathematical, visual/spatial, and naturalist intelligences.

ACTIVITY 1

The Bowling Game

MATH TOPICS

Computation, Order of Operations, Problem Solving

TYPES OF INTELLIGENCES

Logical/Mathematical, Interpersonal, Intrapersonal, Verbal/Linguistic, Musical/Rhythmic

CONCEPTS

Students will do the following:

1. Use order of operations to calculate the numbers 1 through 10.
2. Practice computational skills.

MATERIALS

- Copy of Game Sheet for each player for each game
- Copy of Grading Matrix for each group
- One die for each player
- Timer or watch
- Copy of "O^3—Order of Operations Song" for each student (optional)

WHAT TO DO

Review order of operations with students. Remind them to first calculate what appears inside of parentheses, then to compute exponents and square roots, then to divide and/or multiply (as numbers appear from left to right), and finally to add and/or subtract (as numbers appear from left to right). One of the music medleys, "O^3—Order of Operations Song" (see appendix), is perfect to review the rules with students.

Students can play in pairs or groups of four. Each player needs his or her own game sheet. Players need one game sheet for each game they play. A maximum of 5 minutes has been allotted for each game, however, the time can be changed to suit the needs and skills of your class. The rules are very simple:

1. Each player rolls one die 4 times and enters the numbers in the spaces provided. Each of the players should be playing with his or her own set of 4 numbers.

2. When all players have acquired their numbers, the timing begins. Each player has 5 minutes to find as many of the numbers (1–10) as possible with his or her set of 4 numbers. When a number is found, the corresponding bowling pin is crossed off, or "knocked down." Players can add, subtract, multiply, divide, use exponents or square roots, or use any combination of the above. *But they must follow the order of operations!*

3. The players receive 1 point for each number found, but if all 10 numbers are found, they receive 20 points because they have a strike.

4. When time is up, the teams can play another game. Each player gets to roll the die a second time, recording the new set of 4 numbers.

5. Play continues until one of the players has 30 points or until time has expired for the round of play (to be determined by the teacher).

Students are given space on the game sheet to record their solutions so these sheets can be examined and assessed. It is important that students show their work so their understanding can be evaluated.

VARIATION

Any set of numbers can be used. Rather than selecting numbers by rolling a die, specific numbers can be assigned; for example, to celebrate Independence Day, students can be asked to use the numbers 1, 7, 7, 6 to find the numbers 1–10. They can use the month, day, and last two digits of the year for a specific date; for example, May 1, 1999 would be the digits 5, 1, 9, 9. Be creative! Let the students help you find unique number combinations for this activity!

TIE TO TECHNOLOGY

How many different combinations of numbers are available to the student with each game they play? For example, if the numbers are 2, 6, 1, and 4, what combinations are available? Scientific calculators can be used to find the number of combinations and then students can, using pencil and paper or computer programs, find all of the possible combinations.

ASSESSMENT

1. Traditional test or quiz that focuses on order of operations.
2. Student products—game sheets can be evaluated.
3. Journal question: "Explain why order of operations is important when solving problems that contain more than one mathematics operation. Give an example of a problem where different answers are possible if one does not correctly follow the order of operations."
4. Grading matrix

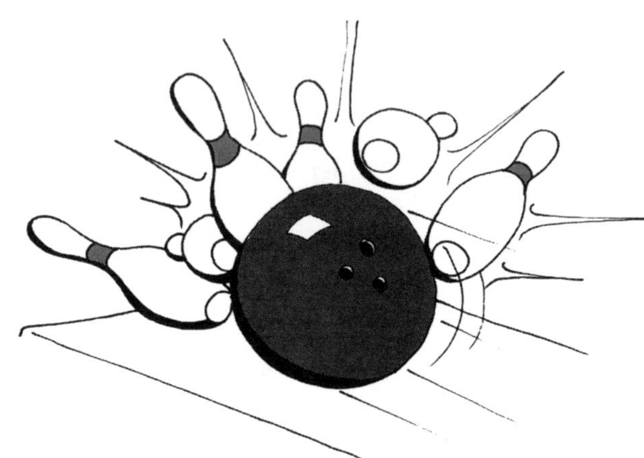

The Bowling Game

Game Sheet

Directions

Roll one die 4 times. Record each number in the spaces below. Using your set of 4 numbers and the order of operations, find as many numbers between 1 and 10 as possible. Each time you find a number, "knock down" the corresponding bowling pin. You get 1 point for each pin you knock down. If you find all 10 numbers, you have a strike and get a total of 20 points. The game continues for 5 minutes or until no player can find any more numbers (whichever comes first). Then, each player rolls the die again and begins a new game. The first player with 30 points wins. Be sure to show your work for each game in the space provided.

My Numbers | | | |

7 8 9 10

4 5 6

2 3

1

My Solutions

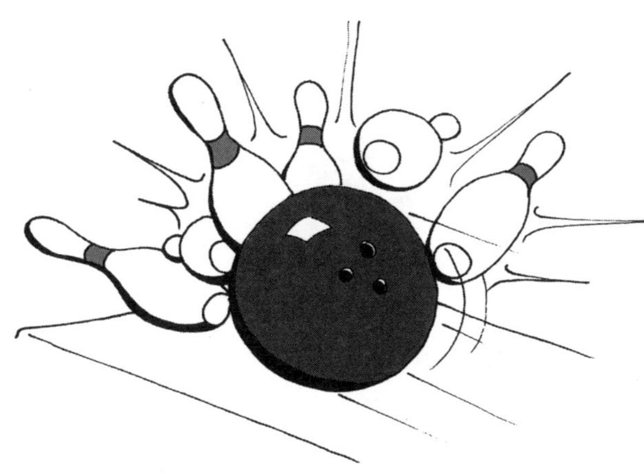

The Bowling Game

Grading Matrix

Names ____

Date __________________ Class ___________________________

Criteria	**4**	**3**	**2**	**1**
Accuracy of computation				
Correct use of order of operations				
How many solutions were found?				
Participation in group				

Comments:

4 = Superlative, 3 = Competent, 2 = Limited, 1 = Inadequate

ACTIVITY 2

All About the Moon

MATH TOPICS

Computation, Rounding, Open-Ended Problem Solving, Conversions

TYPES OF INTELLIGENCES

Logical/Mathematical, Verbal/Linguistic, Interpersonal

CONCEPTS

Students will do the following:

1. Solve problems using very large numbers.
2. Convert units of measurement.
3. Round numbers to appropriate place values.
4. Use reference materials to understand large numbers.

MATERIALS

- Copy of Activity Sheet for each student
- Copy of Grading Matrix for each student
- World atlas for each student
- Calculators

WHAT TO DO

Explain to students that this activity includes some very interesting facts about Earth's moon and asks us to try to make sense of some enormous numbers by comparing them to things we do in our everyday lives. For example, can we really understand how far 239,000 miles is? It doesn't seem like a very large number, but traveling at 75 mph it would take a person 3,186 $^{2}/_{3}$ days or about 8 years and 267 days to travel this distance. When students relate large numbers to familiar topics, they start to believe that "math really does make sense!"

Distribute a copy of Activity Sheet to each student. Students will come up with various answers to each question depending on how they round out numbers. Following are *possible* answers to each of the activity sheet questions:

1. Some countries or states that fit the criteria:
   The crater is about 4,000 square miles larger than Croatia and almost 60 times larger than the island Martinique. It is just slightly smaller than the size of South Carolina but more than three times the size of New Jersey.

2. It would take 3,186$^{2}/_{3}$ days, or 8 years, 267 days, to take the train ride to the moon.

3. The moon rock cost $30,000,000 per pound or $1,875,000 per ounce.

Give students an opportunity to explain not only what answers they came up with but how they solved the problems. By doing this, you are saying the *process* (thinking) is as valuable as the *product* (answer).

VARIATION

Students can problem solve the following questions:

1. How many students would fit in the crater?

2. How long would it take to fly to the moon (at a speed of 500 mph)?

3. How much per ounce does a sneaker or math book cost?

TIE TO TECHNOLOGY

The second problem asks students to calculate the amount of time it would take them to reach the moon if they traveled at a rate of 75 mph. Using a graphing calculator and the equation $\frac{d}{r}=t$, students can locate any coordinate on the linear graph produced by the calculator and determine the distance they traveled at any time during their $8^2/_3$-year journey.

ASSESSMENT

1. Student Activity Sheet
2. Journal question: "You are told that someone traveled 31,680,000 inches from her home to New York. What is wrong with the way this information is presented? What might you do to correct it? What would be a more reasonable way to express your answer? Explain how you solved this problem."
3. Grading matrix

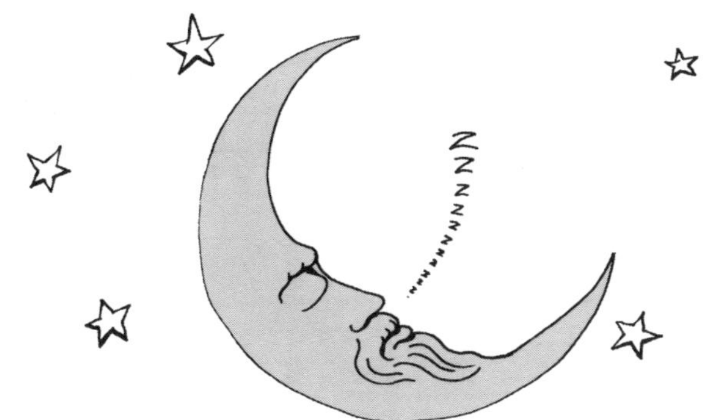

All About the Moon

Activity Sheet

1. Did You Know?

The largest crater we can see on the moon is called Bailly and covers an area of about 26,000 square miles.

Problem

Use a U.S. or world atlas to find what states or countries could fit into a crater of this size.

2. Did You Know?

The average distance of the moon from Earth is 239,000 miles.

Problem

If you were traveling on a train going at a speed of 75 mph, how long would it take you to reach the moon? (Be sure to use appropriate units of measure.)

3. Did You Know?

The Apollo Space Program has cost about $25,500,000,000. The Apollo astronauts who landed on the moon brought about 850 pounds of rocks back to Earth.

Problem

About how much did we spend for each pound of moon rock?
. . . for each ounce of moon rock?

All About the Moon

Grading Matrix

Names ______________________________

Date__________________ Class ______________________________

Criteria	**4**	**3**	**2**	**1**
Accuracy of solution—square miles				
Accuracy of solution—traveling by train				
Accuracy of solution—cost of moon rock				
Overall quality of work				
Comments:				

4 = Superlative, 3 = Competent, 2 = Limited, 1 = Inadequate

ACTIVITY 3

Cooking and Mathematics: Converting Recipes

MATH TOPICS

Ratio/Proportion, Rational Numbers, Computation, Estimation, Rounding, Problem Solving

TYPES OF INTELLIGENCES

Logical/Mathematical, Interpersonal, Bodily/Kinesthetic, Verbal/Linguistic, Intrapersonal

CONCEPTS

Students will do the following:

1. Calculate a conversion ratio.
2. Use that ratio to convert a recipe to make the suitable number of cookies.
3. Round their conversions to appropriate measurements.

MATERIALS

- Copy of Recipe Sheet for each student or group
- Copy of Worksheet for each student or group
- Copy of Grading Matrix for each group
- Calculators (if needed)

WHAT TO DO

Ask students if they have ever used mathematics when they were cooking or baking. They might talk about how they needed to convert a recipe to change the number of portions it makes. If not, suggest that this is one way cooks use mathematics. Discuss how professional chefs and bakers have ways to make the conversion of recipes easier and more accurate, as demonstrated in the following method.

Give an example of converting a recipe that serves 6 to one that serves 8. The conversion ratio is

$$\frac{\text{new yield}}{\text{old yield}};\ \frac{8}{6} = 1^1/_3 \text{ or } 1.3$$

This means that there will need to be $1^1/_3$ more of every ingredient in the recipe. But what if the recipe calls for 1 egg—one cannot put in $1^1/_3$ eggs! So what would we do? Put in 1 egg or 2 eggs? Different groups may problem solve different solutions to this problem. One group might put in 1 egg and a little bit more liquid (if it is needed); another might put in 2 eggs and add a tiny bit more flour. This dilemma makes for some interesting alternatives! The Pecan Crispies recipe in this activity makes 24 cookies and students are asked to convert it so each student gets 3 cookies. Therefore, the conversion ratio for your class will be

$$\frac{\text{the number of students in class divided}}{24} \times 3$$

VARIATIONS

1. Students can be asked to bring in a recipe (perhaps a family favorite) and convert it so that it can be used to make enough food for every student in the class.

2. Students can be asked to convert the cookie recipe so that it makes enough cookies for everyone in the entire grade or perhaps the entire school.

3. Students can research the population of the town or city and convert the cookie recipe to make enough for the entire town. For this question, some additional information is helpful:

4 cups of flour = 1 pound

2.5 cups of brown sugar = 1 pound

2 cups of butter = 1 pound

3 teaspoons = 1 tablespoon

4 tablespoons = $^1/_4$ cup

TIE TO TECHNOLOGY

Spreadsheets can be used to convert any recipe to feed the number of people in the student's family. An example of a spreadsheet is shown below. It assumes the student has 5 people in her family, and the recipe is written to feed 6 people. The conversion ratio is 0.83. The second table shows the formulas entered into the spreadsheet. If one entered in a recipe that only fed 4 people, the spreadsheet would automatically change to a conversion ratio of 1.25 and all of the other quantities would change accordingly.

	A	B	C
1	# of People in Family	# of Servings in Recipe	Conversion Ratio
2	5.00	6.00	0.83
3	Ingredients		Amount Needed
4	flour	4.00	3.33
5	sugar	2.00	1.67
6	butter	0.50	0.42
7	milk	1.50	1.25

	A	B	C
1	# of People in Family	# of Servings in Recipe	Conversion Ratio
2	5	6	=A2/B2
3	Ingredients		Amount Needed
4	flour	4	=B4*C$2
5	sugar	2	=B5*C$2
6	butter	0.5	=B6*C$2
7	milk	1.5	=B7*C$2

ASSESSMENT

1. Observation and questioning: "Can you explain how you decided how many eggs to put in the new recipe?" or "When you found it necessary to approximate a quantity, how did you decide what was the most appropriate way to round it?"

2. Journal question: "Explain how conversion ratios are used with recipes. Give an example of how and when you might use one."

3. Grading matrix

ON THE INTERNET

Students can obtain recipes for practically every imaginable food by searching the Internet. Entering a search term such as "cookie recipe" into a search mechanism such as Metacrawler (see appendix for URL) will pull up a list of numerous sites offering cookie recipes. Some of these sites, such as CookieRecipe.com (see appendix), provide tips on related topics such as shipping cookies, measurements, and conversions, and offer a recipe exchange.

Cooking and Mathematics: Converting Recipes

Recipe Sheet

People use a lot of mathematics when they cook. Sometimes a person needs to convert a recipe because it doesn't yield the amount of servings needed for the number of people who will be eating. For example, if the recipe is designed to feed 4 people and someone is having 10 people for dinner, the recipe will need to be converted. It easily can be converted by using a conversion ratio. The amount needed is referred to as the yield. The following equation is used to determine the conversion ratio:

$$\frac{\text{new yield}}{\text{old yield}} = \text{conversion ratio}$$

To convert a yield of 4 portions to a yield of 10 portions, one would use the following mathematical equation:

$$\frac{10}{4} = 2.5$$

Everything in the recipe would need to be multiplied by 2.5 to make a large enough quantity to feed 10 people. This method (conversion ratio) can be used to convert the following cookie recipe to make enough cookies so that everyone in the class can have 3 cookies.

Pecan Crispies
(The original recipe)

1 cup brown sugar (packed)
2 tablespoons flour
1 egg
1 tablespoon butter
1 cup finely chopped pecans
$^{1}/_{4}$ teaspoon salt
1 teaspoon vanilla

Beat together sugar, egg, and butter. Add vanilla, salt, flour, and pecans. Drop by teaspoonfuls on greased and floured cookie sheet. Bake at 350° for 10 minutes. Makes 2 dozen cookies.

Cooking and Mathematics: Converting Recipes

Worksheet

How many cookies will you need to bake to have enough cookies so everyone in the class can have three? ____ The original recipe makes 2 dozen cookies.

$$\frac{\text{new yield}}{\text{old yield}} = \text{conversion ratio}$$

Our conversion ratio is ______

Work with your group to convert the recipe so you have enough cookies. It may be necessary for you to round out some of the ingredients but be very careful! When you change the amount of something (even slightly), you will change the way the final product (cookie) tastes! You will need some additional information to help make your new measurements more appropriate:

4 tablespoons = $^1/_4$ cup

3 teaspoons = 1 tablespoon

Pecan Crispies
(Our converted recipe)

___ cup brown sugar (packed)
___ tablespoons flour
___ egg
___ tablespoon butter
___ cup finely chopped pecans
___ teaspoon salt
___ teaspoon vanilla

Beat together sugar, egg, and butter. Add vanilla, salt, flour, and pecans. Drop by teaspoonfuls on greased and floured cookie sheets. Bake at 350° for 10 minutes. Makes _____ cookies.

Cooking and Mathematics: Converting Recipes

Grading Matrix

Names ____

Date____________________ Class ____________________________________

Criteria	**4**	**3**	**2**	**1**
Accuracy of computation				
Appropriateness of rounded quantities				
Quality of problem solving				
How well group worked together				
Comments:				

4 = Superlative, 3 = Competent, 2 = Limited, 1 = Inadequate

ACTIVITY 4

Target Game

MATH TOPICS

Estimation, Computation, Decimal Numbers

TYPES OF INTELLIGENCES

Logical/Mathematical, Interpersonal, Intrapersonal

CONCEPTS

Students will do the following:

1. Work collaboratively to choose a factor that brings their team within the target range.
2. Use estimation skills to practice multiplication of decimal numbers.

MATERIALS

- Copy of Activity Sheet for each student
- Overhead transparency of "Target Game" board
- Number tiles (15–99) in a bag or other container. (Note: If you wish to show these on the overhead projector, make an overhead transparency of the numbers before cutting them out.)
- Two boxes: Number Picked box and Target Range box
- Calculators

WHAT TO DO

Divide the class into two teams. Explain to students the rules of the game: A number is drawn from the bag. It is placed in the Number Picked box. Then you, as the teacher, choose a range of numbers for the target. The "number picked" is automatically one of the first factors; the team chooses the second factor (one it believes will result in a product *within the target range)*.

Steps:

1. A number is drawn.

2. The teacher arbitrarily chooses a target range; for example, suppose the number drawn is 32 and the teacher selects a target range of 2,200 to 2,300. Following are possible options the teams might choose.

    - Team 1: Must use the number 32 and pick a second factor that will produce a product between 2,200 and 2,300. It chooses 80 as its factor. But 80 × 32 = 2,560. This estimate was too large. Since 2,560 is too large, team 2 needs to choose a factor that is less than 1 to come up with a smaller product.

    - Team 2: Chooses 0.6 as its factor. It multiplies the first product by 0.6, which equals 1,536 (2,560 × 0.6 = 1,536). The product is less than 2,200. This estimate is too small. Since the product is now too small, team 1 has to multiple by a number between 1 and 2 to increase the product.

    - Team 1: Chooses 1.2 as its new factor and multiplies the second product (1,536) by 1.2: 1,536 × 1.2 = 1,843.2. This estimate is still too small. (1,843.2 is less than 2,200.)

    - Team 2: Also chooses 1.2 as its factor: 1,843.2 × 1.2 = 2,211.84. This estimate is within the target range and so team 2 gets 1 point.

At the end of the game, the team with the most points is the winner. Or the game can continue until one of the teams has 10 points.

VARIATION

With young children, the game can be played using addition. With more advanced students, the students can estimate irrational numbers. Any form of computation skill can be practiced using this game.

TIE TO TECHNOLOGY

The NCTM stated that students ". . . should be able to decide when they need to calculate . . . [what is] the most appropriate tool" (1989, 8). When an exact answer is needed, three possible options are available: pencil and paper, calculators, or computers. Students need relevant experiences to help them choose the appropriate tool. The Target Game lends itself to the use of calculators because of the complex and varied calculations required. It is an excellent example of when the use of calculators is appropriate.

ASSESSMENT

1. Informal observation of participants
2. Traditional assessment technique, such as quiz or test

Target Game

Activity Sheet

The object of this game is to see which team can find two factors that will produce a product within a target range first.

Directions

1. The teacher draws a number from 15 to 99 and chooses a target range.
2. Teams take turns multiplying first the selected number and then the previous product by a whole number or a decimal number to come up with an end product, or number, within the target range.
3. The first team to come up with a number within the target range gets 1 point. If the target range is reached on the first or second try, the team receives 2 points.
4. For example: If the teacher draws a 24 from the number tiles and chooses a target range between 4,000 and 4,100, the 24 is placed in the Number Picked box and the range is place in the Target Range box. Teams take turns performing multiplications to reach a number within the target range.
   - Team 1 multiplies by 150; 150 × 24 = 3,600—too small.
   - Team 2 multiplies 3,600 by 1.4; 3,600 × 1.4 = 5,040—too big.
   - Team 1 multiplies 5,040 by 0.8; 5,040 × 0.8 = 4,032—on target.
5. Calculators cannot be used to estimate your guess, only to find the product! Once the factor is chosen, calculators can be used to perform the multiplication. Calculators can not be used to divide to find the factor.

Number Picked: ______

Target Range: ______

Team 1	Team 2

Target Game

Number Tiles

Directions

Cut out these tiles and use them to draw the first factor.

15	16	17	18	19	20	21	22	23	24
25	26	27	28	29	30	31	32	33	34
35	36	37	38	39	40	41	42	43	44
45	46	47	48	49	50	51	52	53	54
55	56	57	58	59	60	61	62	63	64
65	66	67	68	69	70	71	72	73	74
75	76	77	78	79	80	81	82	83	84
85	86	87	88	89	90	91	92	93	94
95	96	97	98	99					

ACTIVITY 5

The Pattern Tells It All

MATH TOPICS

Multiples, Patterns, Critical Thinking, Problem Solving

TYPES OF INTELLIGENCES

Visual/Spatial, Logical/Mathematical, Intrapersonal, Verbal/Linguistic, Naturalist

CONCEPTS

Students will do the following:

1. Shade the squares in a 100s chart using the multiples of the numbers 2, 3, 4, 5, 6, 8, and 9.
2. Discover the visual clues or patterns formed by the number patterns.
3. Describe the pattern formed by prime and composite numbers.

MATERIALS

- Copy of Number Pattern Sheet for each student (each student will need 8 grids)
- Copy of Worksheet for each student
- Copy of Grading Matrix for each student
- Colored pencils or markers

WHAT TO DO

Give each student a copy of the Number Pattern Sheet and ask how it differs from the usual 100s chart. Students should notice that the first row and column has the multiples of 1, the second row and column the multiples of 2, etc., so that numbers are repeated throughout the chart. On the 100s chart, shading in the even-number columns can be used to identify the multiples of 2. On this chart, even numbers can be found throughout and students will need to search for them. The following grid shows the multiples of 3.

1	2	3	4	5	6	7	8	9	10
2	4	6	8	10	12	14	16	18	20
3	6	9	12	15	18	21	24	27	30
4	8	12	16	20	24	28	32	36	40
5	10	15	20	25	30	35	40	45	50
6	12	18	24	30	36	42	48	54	60
7	14	21	28	35	42	49	56	63	70
8	16	24	32	40	48	56	64	72	80
9	18	27	36	45	54	63	72	81	90
10	20	30	40	50	60	70	80	90	100

This grid shows the multiples of 4.

1	2	3	4	5	6	7	8	9	10
2	4	6	8	10	12	14	16	18	20
3	6	9	12	15	18	21	24	27	30
4	8	12	16	20	24	28	32	36	40
5	10	15	20	25	30	35	40	45	50
6	12	18	24	30	36	42	48	54	60
7	14	21	28	35	42	49	56	63	70
8	16	24	32	40	48	56	64	72	80
9	18	27	36	45	54	63	72	81	90
10	20	30	40	50	60	70	80	90	100

Notice that each of the designs is a type of checkerboard, but the multiples of 4 have additional squares in the center of each larger square as part of the pattern, or design. As students color in their designs, they should begin to see this pattern emerge—the prime numbers are exact checkerboards and the composite numbers have additional factors (or shaded squares).

VARIATION

Students can design larger grids to expand the pattern.

TIE TO TECHNOLOGY

The following program (written in BASIC) will produce a list of prime numbers between 2 and 50.

```
10 PRINT "PRIMES"

20 FOR N = 2 TO 50

30 LET NF = 0

40 FOR I = 1 TO SQR(N)

50 IF INT(N/I) = N/I THEN NF = NF + 1

60 NEXT I

70 IF NF = 1 THEN PRINT N;

80 NEXT N

90 END
```

ASSESSMENT

1. Student products (number pattern designs)
2. Journal question: "Explain how the difference between prime and composite numbers is reflected in the grid designs. Describe two numbers that would create each of these patterns."
3. Grading matrix

ON THE INTERNET

One of the most comprehensive mathematics websites on the Internet is The Math Forum (see appendix for URL). It is operated by Swarthmore College and its content frequently changes. However, it always includes information on projects and activities related to prime and composite numbers. Its archives includes a discussion of the use of Mersenne primes (to find the largest prime number).

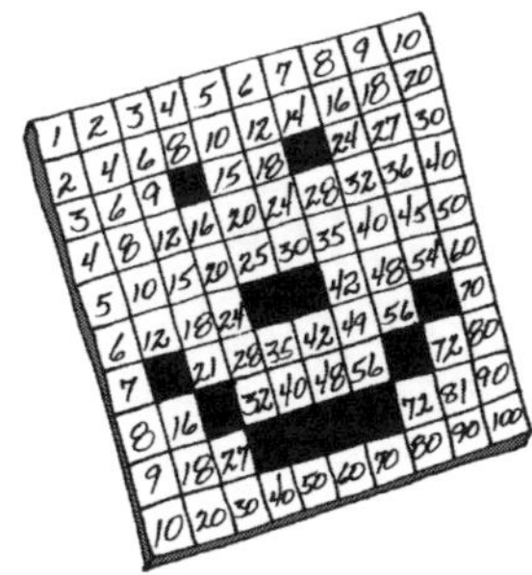

The Pattern Tells It All

Number Pattern Sheet

Directions

To complete this activity, you will need eight number grids. On the first, you will shade all the multiples of 2, on the second the multiples of 3, then the multiples of 4, 5, 6, 7, 8, and 9. When you complete all eight designs, try to find a relationship between the number and the pattern it forms.

1	2	3	4	5	6	7	8	9	10
2	4	6	8	10	12	14	16	18	20
3	6	9	12	15	18	21	24	27	30
4	8	12	16	20	24	28	32	36	40
5	10	15	20	25	30	35	40	45	50
6	12	18	24	30	36	42	48	54	60
7	14	21	28	35	42	49	56	63	70
8	16	24	32	40	48	56	64	72	80
9	18	27	36	45	54	63	72	81	90
10	20	30	40	50	60	70	80	90	100

1	2	3	4	5	6	7	8	9	10
2	4	6	8	10	12	14	16	18	20
3	6	9	12	15	18	21	24	27	30
4	8	12	16	20	24	28	32	36	40
5	10	15	20	25	30	35	40	45	50
6	12	18	24	30	36	42	48	54	60
7	14	21	28	35	42	49	56	63	70
8	16	24	32	40	48	56	64	72	80
9	18	27	36	45	54	63	72	81	90
10	20	30	40	50	60	70	80	90	100

1	2	3	4	5	6	7	8	9	10
2	4	6	8	10	12	14	16	18	20
3	6	9	12	15	18	21	24	27	30
4	8	12	16	20	24	28	32	36	40
5	10	15	20	25	30	35	40	45	50
6	12	18	24	30	36	42	48	54	60
7	14	21	28	35	42	49	56	63	70
8	16	24	32	40	48	56	64	72	80
9	18	27	36	45	54	63	72	81	90
10	20	30	40	50	60	70	80	90	100

1	2	3	4	5	6	7	8	9	10
2	4	6	8	10	12	14	16	18	20
3	6	9	12	15	18	21	24	27	30
4	8	12	16	20	24	28	32	36	40
5	10	15	20	25	30	35	40	45	50
6	12	18	24	30	36	42	48	54	60
7	14	21	28	35	42	49	56	63	70
8	16	24	32	40	48	56	64	72	80
9	18	27	36	45	54	63	72	81	90
10	20	30	40	50	60	70	80	90	100

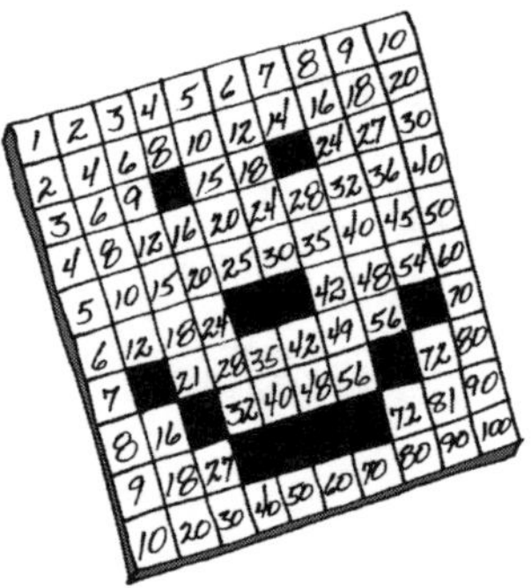

The Pattern Tells It All

Worksheet

Explain the relationship you can see between the numbers you used and the patterns they formed. (Hint: Think about what you know about prime and composite numbers.) Make a prediction—what do you think the pattern for the number 17 might look like; the pattern for the number 24? Explain your reasoning.

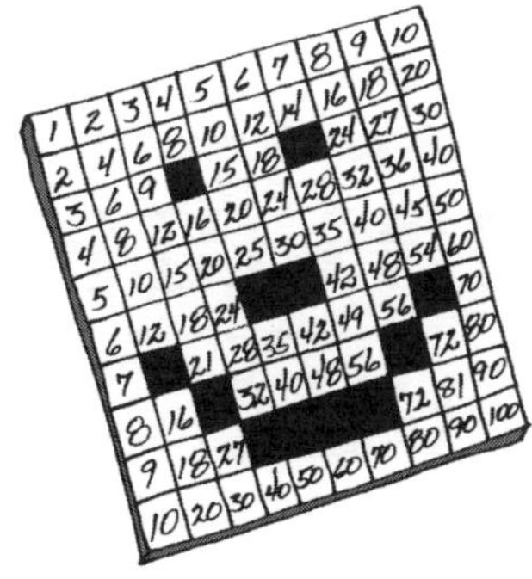

The Pattern Tells It All

Grading Matrix

Names ____

Date____________________ Class ______________________________

Criteria	**4**	**3**	**2**	**1**
Accuracy of patterns				
Overall quality of pattern designs				
Quality of written description				
Comments:				

4 = Superlative, 3 = Competent, 2 = Limited, 1 = Inadequate

ACTIVITY 6

Eratosthenes and the 500 Chart

MATH TOPICS

Prime and Composite Numbers, Divisibility Rules, Computation, Problem Solving

TYPES OF INTELLIGENCES

Logical/Mathematical, Visual/Spatial, Verbal/Linguistic, Musical/Rhythmic

CONCEPTS

Students will do the following:

1. Use divisibility rules to find the multiples of prime numbers.
2. Find the prime numbers ≤ 500.
3. Highlight the prime numbers.

MATERIALS

- Copy of 500 Chart for each student
- Highlighter for each student
- Calculators (if necessary)
- Copy of the song "The Divisibility Ditty" for each student (optional)

WHAT TO DO

Explain that Eratosthenes, a Greek geographer, devised a sieve that he used to separate prime from composite numbers. This activity is an extended version of the technique that he used.

Begin the lesson by reviewing the rules of divisibility with students. This is a good time to use the song "The Divisibility Ditty" (see appendix). It reviews the divisibility rules for the numbers 2, 3, 4, 5, 6, 8, 9, and 10.

Explain to students that they will be finding all of the prime numbers that are ≤ 500. Explain the procedure:

1. Cross out the number 1—it is neither prime nor composite.

2. Highlight the number 2—it is the first prime number. Cross out all the multiples of 2. Ask students, "How will you know if a number is a multiple of two?" Students should know that the multiples of 2 are all the even numbers, or they could count every second number following 2. Give students time to complete this procedure. Ask students, "What percentage of the numbers on the chart have been eliminated when we removed the multiples of two? Did you notice any particular pattern?"

3. Highlight the number 3—it is the next prime number. Now the students must cross out all the multiples of 3. Ask, "Are some of the multiples of three already crossed out? How did this happen? Are any of the multiples of three still remaining? How can we find these?" Suggest that every sixth number (following 3) will be a multiple of 3 still remaining or that they use the divisibility rule: find the sum of the digits, if the sum is divisible by 3, then the number is divisible by 3. For example, using the number 456, 4 + 5 + 6 = 15; 15 is divisible by 3, and so 456 is divisible by 3. Give students time to complete this procedure. Ask, "Did you notice a particular pattern when you eliminated the multiples of three? What method(s) did you use to be sure that you had eliminated the right numbers?"

These two questions can be asked after each step.

4. Ask students what the next prime number is; it's the number 5. Students must highlight this number and cross out all of its multiples. Ask, "How will you know if a number is divisible by five? Are any of the multiples of five crossed out? Why?" Students should now eliminate all of the multiples of 5. (Multiples of 5 end in 5 or 0.) Give students time to complete this procedure.

5. The next prime number is 7. Ask students, "What multiples of seven have been crossed out? What is the first multiple of seven that has not been eliminated?" (7×7 is the first; 7×8, 7×9, 7×10 are also gone; they were eliminated with the multiples of 2 and 3.) The next multiple of 7 to be eliminated is 7×11, then 7×13, then 7×17, etc. Students need only eliminate the multiples of the remaining prime numbers. *The last composite number remaining on the chart will be 19×23. All the rest of the numbers will be prime!*

Following is a list of all of primes ≤ 500.

2	3	5	7	11	13	17	19	23	29
31	37	41	43	47	53	59	61	67	71
73	79	83	89	97	101	103	107	109	113
127	131	137	139	149	151	157	163	167	173
179	181	191	193	197	199	211	223	227	229
233	239	241	251	257	263	269	271	277	281
283	293	307	311	313	317	331	337	347	349
353	359	367	373	379	383	389	397	401	409
419	421	431	433	439	443	449	457	461	463
467	479	487	491	499					

VARIATION

Students can develop a chart of the prime numbers $\leq$ 1,000.

ASSESSMENT

1. Traditional quiz—primes, composites, and divisibility rules
2. Student product—500 Chart
3. Journal question: "Describe all the patterns that you found on the 500 Chart. They can be number patterns or visual (design) patterns. Be as precise as you can."

ON THE INTERNET

The Prime Page website is perhaps the most comprehensive site on the Internet for prime numbers (see appendix for URL). It contains a wealth of information about prime numbers, including the largest known prime, how to find primes, how many primes there are, frequently asked questions about primes, special types of primes, and types of software programs to help find prime numbers.

Students can learn more about Eratosthenes through the University of Utah's website (see appendix for URL). It provides information about the Sieve of Eratosthenes and links to related websites.

500 Chart

Worksheet

Use this chart to find all of the prime numbers ≤ 500. Highlight the prime numbers and cross out all of the multiples of that number. Only primes will remain.

1	2	3	4	5	6	7	8	9	10	11	12	13	14	15	16	17	18	19	20	21	22	23	24	25
26	27	28	29	30	31	32	33	34	35	36	37	38	39	40	41	42	43	44	45	46	47	48	49	50
51	52	53	54	55	56	57	58	59	60	61	62	63	64	65	66	67	68	69	70	71	72	73	74	75
76	77	78	79	80	81	82	83	84	85	86	87	88	89	90	91	92	93	94	95	96	97	98	99	100
101	102	103	104	105	106	107	108	109	110	111	112	113	114	115	116	117	118	119	120	121	122	123	124	125
126	127	128	129	130	131	132	133	134	135	136	137	138	139	140	141	142	143	144	145	146	147	148	149	150
151	152	153	154	155	156	157	158	159	160	161	162	163	164	165	166	167	168	169	170	171	172	173	174	175
176	177	178	179	180	181	182	183	184	185	186	187	188	189	190	191	192	193	194	195	196	197	198	199	200
201	202	203	204	205	206	207	208	209	210	211	212	213	214	215	216	217	218	219	220	221	222	223	224	225
226	227	228	229	230	231	232	233	234	235	236	237	238	239	240	241	242	243	244	245	246	247	248	249	250
251	252	253	254	255	256	257	258	259	260	261	262	263	264	265	266	267	268	269	270	271	272	273	274	275
276	277	278	279	280	281	282	283	284	285	286	287	288	289	290	291	292	293	294	295	296	297	298	299	300
301	302	303	304	305	306	307	308	309	310	311	312	313	314	315	316	317	318	319	320	321	322	323	324	325
326	327	328	329	330	331	332	333	334	335	336	337	338	339	340	341	342	343	344	345	346	347	348	349	350
351	352	353	354	355	356	357	358	359	360	361	362	363	364	365	366	367	368	369	370	371	372	373	374	375
376	377	378	379	380	381	382	383	384	385	386	387	388	389	390	391	392	393	394	395	396	397	398	399	400
401	402	403	404	405	406	407	408	409	410	411	412	413	414	415	416	417	418	419	420	421	422	423	424	425
426	427	428	429	430	431	432	433	434	435	436	437	438	439	440	441	442	443	444	445	446	447	448	449	450
451	452	453	454	455	456	457	458	459	460	461	462	463	464	465	466	467	468	469	470	471	472	473	474	475
476	477	478	479	480	481	482	483	484	485	486	487	488	489	490	491	492	493	494	495	496	497	498	499	500

Math and Music

MATH TOPICS

Fractions, Computation, Problem Solving

TYPES OF INTELLIGENCES

Musical/Rhythmic, Logical/Mathematical, Visual/Spatial

CONCEPTS

Students will do the following:

1. Discover the connections between mathematics and musical composition.
2. Identify musical symbols and see how they relate to rhythms.
3. Learn to apply the musical signature to the rhythm of a musical selection.

MATERIALS

- Copy of Worksheet 1 for each student
- Copy of Worksheet 2 for each student
- Copy of Grading Matrix for each student

WHAT TO DO

Review with students the information in the lesson. Be sure they understand what each symbol represents and how it relates to the musical signature. The duration of each note is relative but a half note will always be given twice the duration as a quarter note. Each of the notes and rests are pictured in the lesson, as well as dotted notes (indicating the value of the note plus half the original value). For example, a dotted quarter note = $^1/_4 + ^1/_8 = ^3/_8$.

Once students understand the examples given, ask them to invent their own song using the information they learned.

VARIATION

Many students have studied music and are quite accomplished. These students can be encouraged to share their expertise with the class, perhaps by beating out rhythms and having the class guess what the time signature might be.

TIE TO TECHNOLOGY

Many computer programs are available that allow users to write music using a computer. Students can research these programs and use the computer to compose their own songs.

ASSESSMENT

1. Student products
2. Observation and questioning: "Can you explain how you determined the time signature for this piece of music?"
3. Journal question: "You come across a piece of music that looks like this. Analyze the beats and determine what the time signature might be. Explain how you got this answer."

4. Grading matrix

Math and Music

Worksheet 1

Music is very mathematical! It is a language consisting of symbols for notes and time. The placement of notes on the musical staff represents their different tones, and the symbol used for the tone indicates its duration or rhythm. The result is a remarkable combination of the sciences of mathematics and physics and the soul of music. The notes shown on this scale all have the same beat but are different tones. They form the C scale. The notes and the keys on the piano used to play those notes are shown.

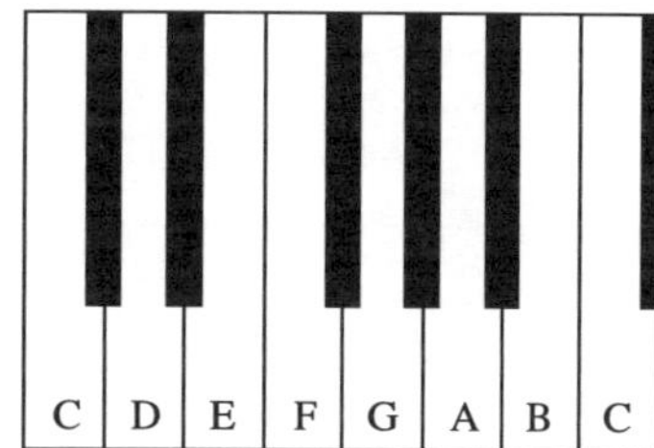

Fractions play an important part of "keeping the beat." Different symbols are used to indicate the duration (or beats) given to a particular note. At the beginning of a musical piece, there is a time signature that looks like a fraction. The bottom number indicates which notes get a count of one beat and the top number indicates how many beats there are in one measure. The time signature $\frac{3}{4}$ signifies that a quarter note gets one beat and there are three beats (or the equivalent of 3 quarter notes) in one measure.

All of these measures are different ways to show the rhythm of three-quarter time.

Let's see what these notes stand for by examining the notes and rests shown below.

These are the symbols used to show the duration of each note and the rests that count time without using additional notes. Sometimes we see a note that is followed by a dot.

This note is called a dotted quarter note and it is the same as $\frac{1}{4}+\frac{1}{8}$, just like the notes show.

Now let's look at each of the $\frac{3}{4}$ measures again and see how the fractions add up.

The first measure is $\frac{1}{2}+\frac{1}{8}+\frac{1}{8}=\frac{6}{8}=\frac{3}{4}$.

You do the next three.

The second measure is ______________________________

The third measure is ______________________________

The fourth measure is ______________________________

How many different measures with a value of $\frac{3}{4}$ can you calculate? Draw your answers on this musical staff.

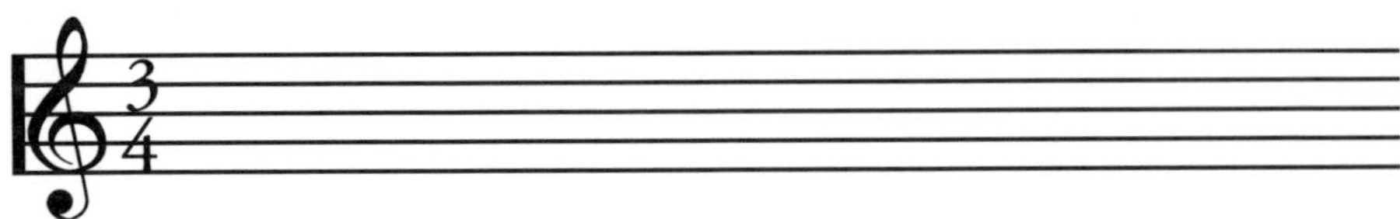

Math and Music

Worksheet 2

Someone has made a terrible mistake—they've printed a great deal of music but left the time signatures off. How will people know what notes get what beats and how many beats there will be in a measure? A mathematician said, "That's not a problem. We can tell what the time signature should be by examining the music. The music holds the clue." See if you can help out the mathematician by analyzing these musical pieces and problem solving what the missing time signatures should be.

Design your own time signature and write 3 bars of music. Use both notes and rests to write your song.

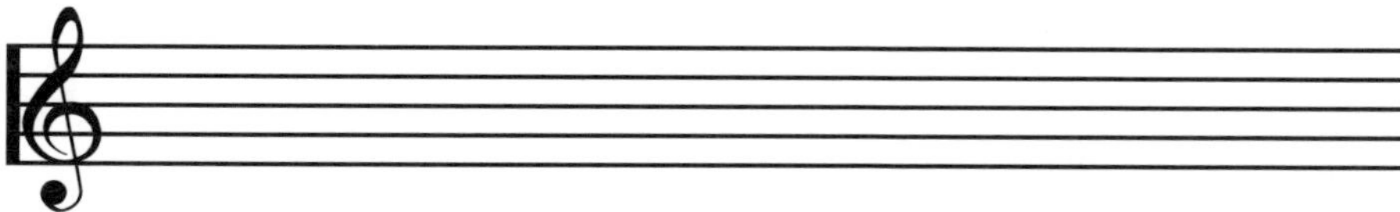

Math and Music

Grading Matrix

Names ____

Date________________ Class ______________________________

Criteria	4	3	2	1
Ability to read musical notes				
Creativity of $\frac{3}{4}$ measures				
Analyzing measures				
Creativity of original composition				
Comments:				

4 = Superlative, 3 = Competent, 2 = Limited, 1 = Inadequate

ACTIVITY 8

Diagramming Divisibility

MATH TOPICS

Divisibility Rules, Multiples, Factors, Venn Diagrams

TYPES OF INTELLIGENCES

Naturalist, Logical/Mathematical, Interpersonal, Intrapersonal

CONCEPTS

Students will do the following:

1. Use a triple-Venn diagram.
2. Use divisibility rules to find the multiples of the numbers 2, 3, and 5.
3. Design their own triple-Venn diagram.

MATERIALS

- Copy of Worksheet 1 for each student
- Copy of Worksheet 2 for each pair
- Copy of Grading Matrix for each pair
- Copy of the musical medley "The Divisibility Ditty" for each student (optional)

WHAT TO DO

Discuss the concept of the triple-Venn diagram with students. Provide an example and walk through it with the students (e.g., students wearing jeans, students wearing red shirts, students with brown hair). Have students classify themselves and problem solve where they belong on this Venn diagram.

Give each student a copy of Worksheet 1. You can use the musical medley "The Divisibility Ditty" (see appendix) to review the divisibility rules before students complete the Venn diagram on Worksheet 1. When individuals have completed the assignment, place them into pairs and give each pair a copy of the triple-Venn sheet (Worksheet 2). Have students work together to create an original Venn diagram. Provide possible ideas for subjects to diagram, such as plant species or animal species, or allow students to generate their own categories to diagram.

The solution to the triple-Venn diagram is shown below.

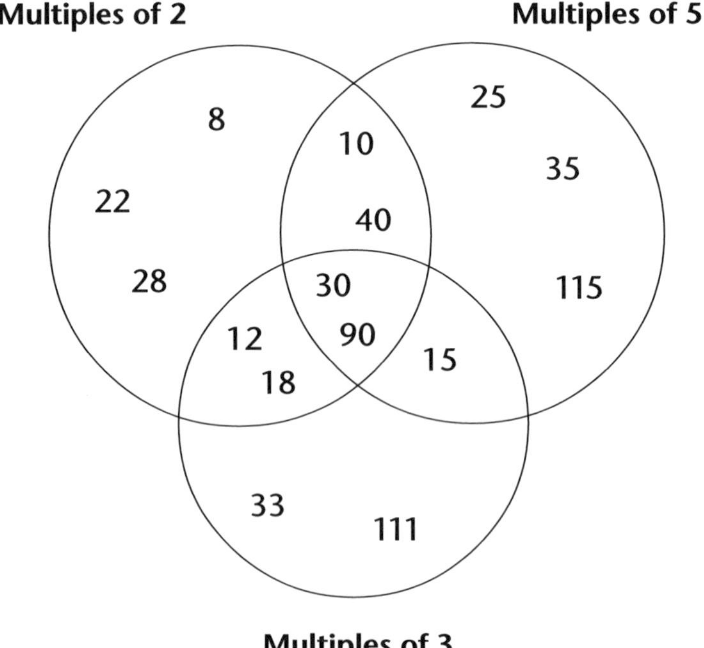

VARIATION

Students can work on a double-Venn, a quadruple-Venn, or other logic-type problems.

TIE TO TECHNOLOGY

This lesson involves students in the exploration of divisibility. The following program (written in BASIC) will print all of the factors of a natural number (N). The program will look like this:

```
10 PRINT "ENTER NUMBER"

20 INPUT N

30 FOR I = 1 TO N

40 IF INT(N/I) = N/I THEN PRINT I

50 NEXT I

60 END
```

Students enter a number and the program will print out all of the natural factors of that number.

ASSESSMENT

1. Student products
2. Observation and questioning: "Can you explain why you placed (a particular number) in that part of the Venn diagram?"
3. Journal question: "You are examining a Venn diagram that depicts the multiples of 2, 5, and 7. There is no number in the center region [where all three circles intersect]. Give an example of a number that might be correctly placed in this region. Explain your answer."
4. Grading matrix

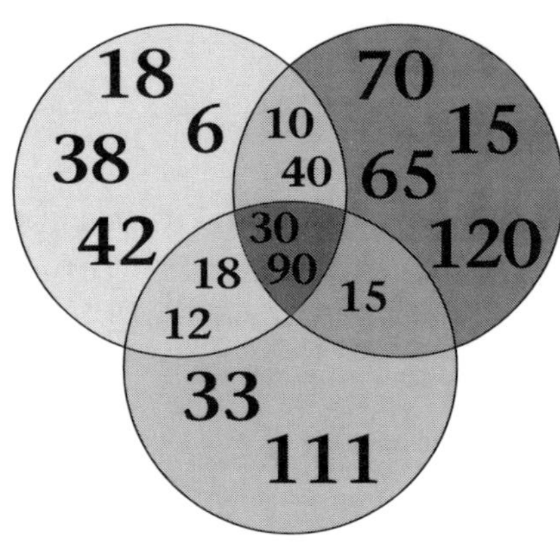

Diagramming Divisibility

Worksheet 1

Directions

Place the numbers at the bottom of the page in the correct circle. If it has something in common with two circles, place it in the space that is common to both; if it has something in common with all three circles, place it in the space where all three circles intersect.

Multiples of 2

Multiples of 5

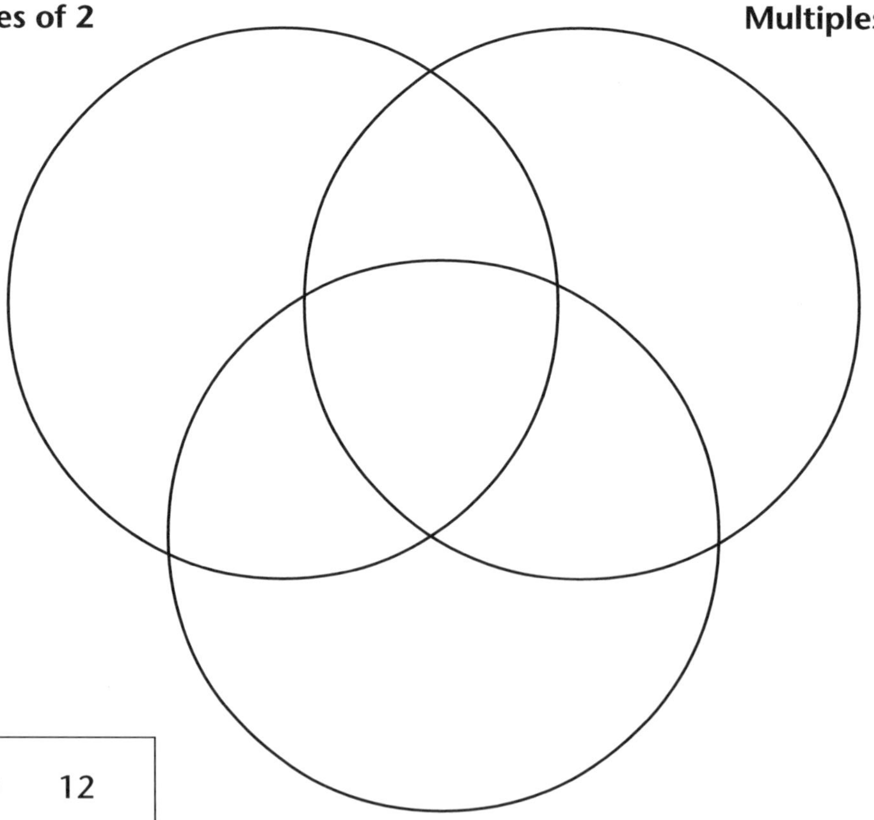

Multiples of 3

8	10	12
15	18	22
25	28	30
33	35	40
90	111	115

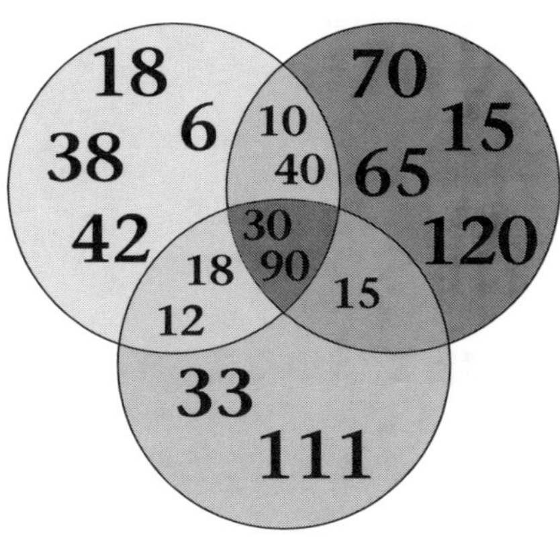

Diagramming Divisibility

Worksheet 2

Names ____

Class__________________ Date______________________________

Work with a partner and create a triple-Venn diagram.

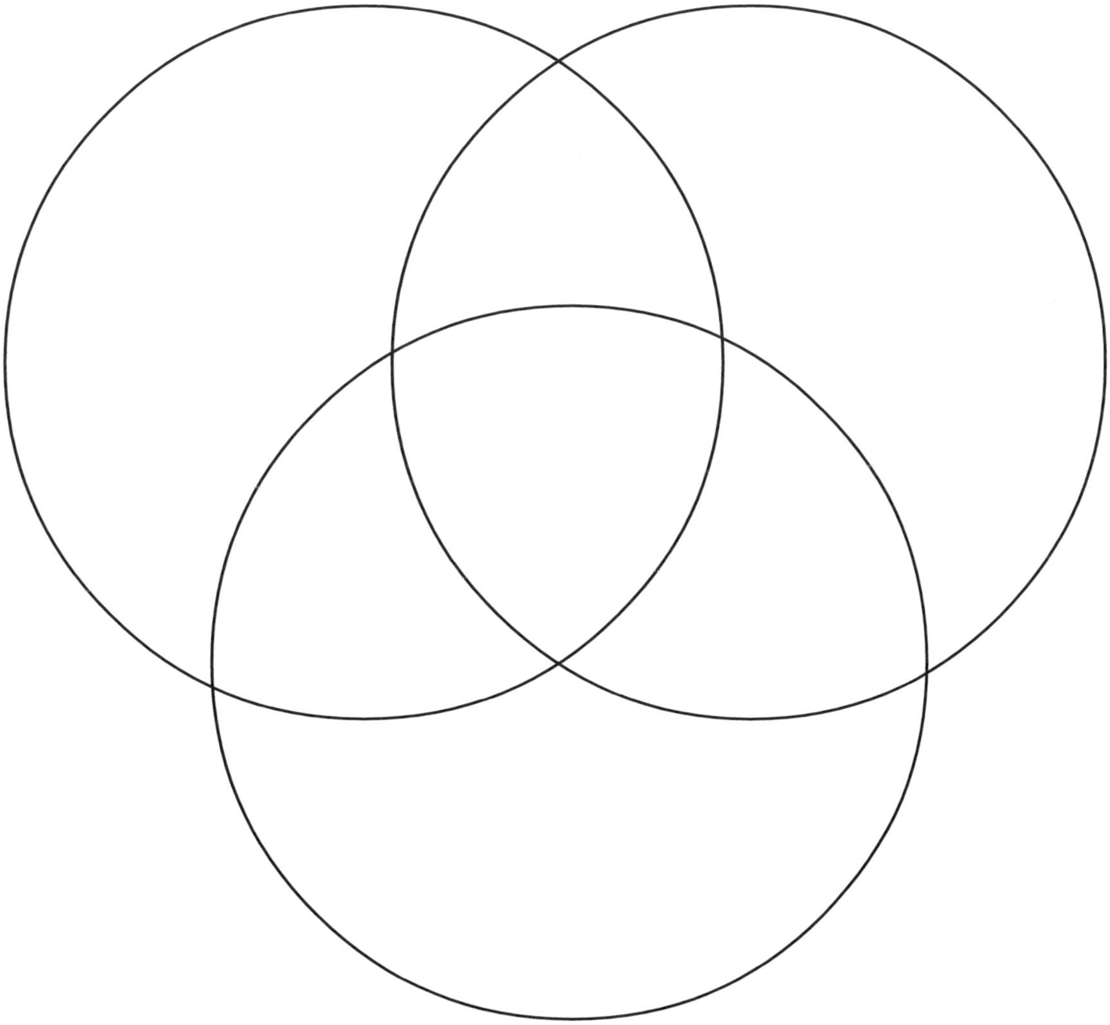

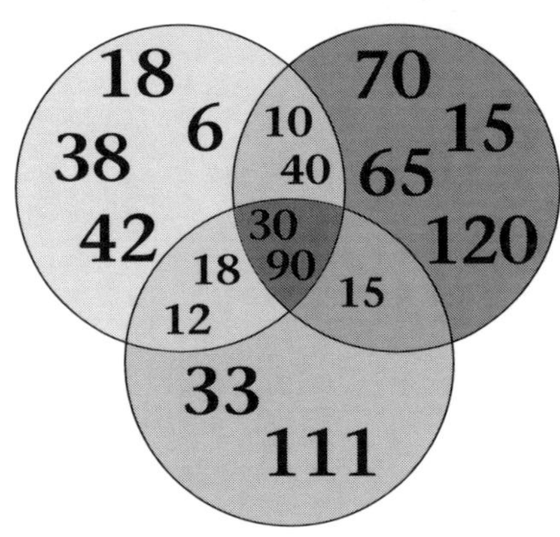

Diagramming Divisibility

Grading Matrix

Names ____

Date__________Class ______________________________

Criteria	**4**	**3**	**2**	**1**
Accuracy of triple-Venn (Divisibility)				
Accuracy of original triple-Venn diagram				
Creativity and originality of original triple-Venn diagram				
How well did pair work together?				
Comments:				

4 = Superlative, 3 = Competent, 2 = Limited, 1 = Inadequate

CHAPTER 2

Making Algebra Come Alive

CHAPTER 2

Making Algebra Come Alive

Mention the word mathematics and one rarely gets an indifferent response . . . the emotions range from fear and hate to love and delight. Some people boast with pride of their ignorance of mathematics. Others go to almost any means to avoid it.—Theoni Pappas 1995, 1

Besides the fear and discomfort that Pappas alludes to, for most students the study of algebra evokes the question, "When am I ever going to use this?" To advance their study of mathematics, students must be able to connect the concrete concepts of arithmetic with the abstract concepts of algebra. However, when algebra is seen as a collection of rules and procedures that make no sense, students do not understand its relevance and lack motivation. The *Curriculum and Evaluation Standards for School Mathematics* (NCTM 1989) expresses the need for students to understand the concept of variable, solve linear equations using concrete activities, and apply algebraic methods to solving real-world mathematical problems. The activities in this chapter allow students to make strong connections between the algebraic patterns and artistic designs and the relationship between algebraic functions (formulas) and real-world phenomena.

Students gain a firm understanding of the Pythagorean Theorem and develop multiple intelligences while carefully measuring the triangles that eventually will form the Irrational Spiral. Before students start this project, they can review the Pythagorean Theorem with the musical ditty "Ode to Pythagoras." After practicing on a standard 22 cm × 28 cm ($8^1/_2$" × 11") sheet of paper, students can, working in a group of four, make a very large spiral that wraps around itself. The more accurate the measurements, the more amazing the spiral.

Formulas in the real world come alive for students as they work on three different activities. The first, Vet Math, asks students to calculate how much fluid a dog would need if it were dehydrated and were going to receive fluid through an intravenous drip. The quantities are realistic, as is the formula, which is the one used by veterinarians to calculate the rate of flow in intravenous drips. After students find the amount of fluid needed for the listed breeds, they can be encouraged to stretch their interpersonal intelligences and survey their classmates and, perhaps, every person in the school to determine what dogs they own. With this amount of data, it would become a real challenge to calculate fluid requirements. Since the calculations are repetitive, students can enter the data into a spreadsheet program and have the computer do their work for them.

Seeing to the Horizon explains the scientific principle that on a clear day, "the higher up you are, the further you can see." A table of ten of the tallest habitable buildings is supplied and students are asked to use a formula to calculate to the nearest mile the distance they can see to the horizon. This is a perfect time to work with the art teacher to introduce perspective and vanishing point to students.

The final activity to bring formulas from the real world into the mathematics classroom is Falling Objects. It asks students to calculate how long it would take for a marshmallow thrown off the top of any of several of the tallest habitable buildings to reach the ground and to determine how fast it was traveling immediately before it reached the ground. Students answer both of these questions by using the formula for vertical drop and acceleration.

When most people think of linear equations, they think of a straight line. But by using modular (or clock) arithmetic, a person can produce wonderful designs within a circle. The 18-Hour Clock: Modular Designs introduces students to a component of finite mathematics while expanding their visual/spatial and logical/mathematical intelligences.

Several of the activities suggest ways to incorporate technology, especially by using spreadsheets. For these activities, sample spreadsheets are shown in the Tie to Technology section of the Teacher's Page.

ACTIVITY 9

The Irrational Spiral

MATH TOPICS

Pythagorean Theorem, Substitution for a Variable, Angle Measurement, Using a Protractor, Irrational Numbers, Problem Solving

TYPES OF INTELLIGENCES

Visual/Spatial, Logical/Mathematical, Interpersonal, Bodily/Kinesthetic, Musical/Rhythmic

CONCEPTS

Students will do the following:

1. Use the Pythagorean Theorem to calculate the lengths of the sides of triangles.
2. Draw these triangles in a spiral to create the irrational spiral.

MATERIALS

- Copy of Activity Sheet for each student
- Copy of Grading Matrix for each pair
- Copy of math medley "Ode to Pythagoras" for each student

- Large sheet of construction paper for each pair
- Protractors, rulers
- Calculators
- Colored pencils

WHAT TO DO

Introduce the Pythagorean Theorem with the math medley "Ode to Pythagoras" (see appendix). Review the theorem and be sure that all of the students understand how to solve the equation and use square roots.

Give each student a copy of Activity Sheet. Have students describe the irrational spiral in their journals and then share their descriptions with the class.

Ask students to find the square root of 3. If available, use an overhead calculator to discuss if the number appears to be a terminating or a repeating decimal. Explain to students that this type of number is called *irrational* and does not terminate or repeat. Have students round the square root of 3 to the nearest hundredth. Go around the picture of the irrational spiral and convert each of the irrational numbers to rational approximations.

Allow students to work in pairs and distribute sheets of construction paper. Have each pair draw an irrational spiral as large as possible on the paper. Tell students that it will start to "wrap around" and they will not be able to continue connecting their lines to the center. Have them stop at the edge of the design so that the lines do not overlap.

It is essential that their measurements be as accurate as possible or the spiral will not look right. They should use the Pythagorean Theorem to find the new side and write that proof down—it should be turned in with the drawing for full credit.

VARIATION

It is possible to cut out each triangle from fabric and sew an irrational spiral quilt. Use one-cm graph paper as a pattern.

Add $^{1}/_{2}$ cm around each line of the triangle for seam allowance. Each triangle can be made about the same size as those on the drawing shown on the activity sheet.

TIE TO TECHNOLOGY

Students can use a computer to discover as many irrational numbers as they like. By using the "fill down" function, they can find the irrational numbers through 1,000 or beyond. Following is a sample spreadsheet showing both numbers and formulas.

	A	B
1	**Number**	**SQRT(N)**
2	1	1
3	2	1.414213562
4	3	1.732050808
5	4	2
6	5	2.236067977
7	6	2.449489743
8	7	2.645751311
9	8	2.828427125
10	9	3
11	10	3.16227766
12	11	3.31662479
13	12	3.464101615
14	13	3.605551275
15	14	3.741657387
16	15	3.872983346
17	16	4
18	17	4.123105626
19	18	4.242640687
20	19	4.358898944

	A	B
1	**Number**	**SQRT(N)**
2	1	=SQRT(A2)
3	=A2+1	=SQRT(A3)
4	=A3+1	=SQRT(A4)
5	=A4+1	=SQRT(A5)
6	=A5+1	=SQRT(A6)
7	=A6+1	=SQRT(A7)
8	=A7+1	=SQRT(A8)
9	=A8+1	=SQRT(A9)
10	=A9+1	=SQRT(A10)
11	=A10+1	=SQRT(A11)
12	=A11+1	=SQRT(A12)
13	=A12+1	=SQRT(A13)
14	=A13+1	=SQRT(A14)
15	=A14+1	=SQRT(A15)
16	=A15+1	=SQRT(A16)
17	=A16+1	=SQRT(A17)
18	=A17+1	=SQRT(A18)
19	=A18+1	=SQRT(A19)
20	=A19+1	=SQRT(A20)

ASSESSMENT

1. Student product (irrational spiral)
2. Observation and questioning: "Can you and your partner draw a line using the rational approximations and have an accurate drawing? Why do you believe this?"
3. Journal question: "Explain the difference between rational and irrational numbers. Be as specific as you can and give an example of each. Be sure to explain why each of these numbers is rational or irrational."
4. Grading matrix

The Irrational Spiral

Activity Sheet

The irrational spiral is formed by careful measurement using a protractor and a metric ruler. Start with a right triangle with sides of 1 cm. When you draw the hypotenuse, you have a length of $\sqrt{2}$. Let's see why this happens.

Using the Pythagorean Theorem, we know that $a^2 + b^2 = c^2$. Since each side is equal to 1 by substitution, we see that $1^2 + 1^2 = c^2$. So $c^2 = 2^2$, or $c = \sqrt{2}$. By using the hypotenuse of this triangle as one of the sides of a new triangle (the other leg remains 1), the length of the new hypotenuse is $1^2 + (\sqrt{2})^2 = c^2$. $1 + 2 = c^2$, $c^2 = 3$, so $c = \sqrt{3}$. An example of this irrational spiral is shown below.

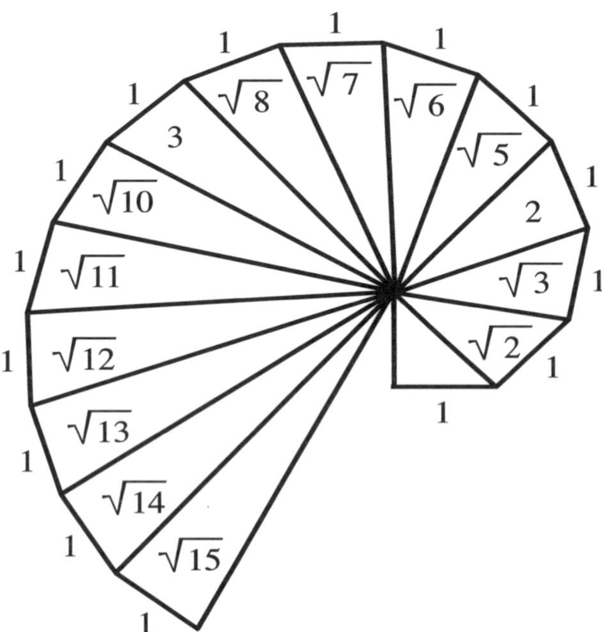

Use your calculator to find the $\sqrt{3}$. What number appears on the display? Does it appear to terminate or repeat? The sides that do not have a whole number (rational root) are called irrational numbers. What irrational numbers are shown in the above illustration?

Some of the sides shown are rational numbers. What will the next rational number length be? Use a large piece of paper and see how large an irrational spiral you can design!

The Irrational Spiral

Grading Matrix

Names ____

Date____________________ Class ______________________________

Criteria	**4**	**3**	**2**	**1**
Accuracy of drawing				
Completeness of written proofs				
Overall quality of design				
Comments:				

4 = Superlative, 3 = Competent, 2 = Limited, 1 = Inadequate

ACTIVITY 10

Vet Math: Formulas in the Real World

MATH TOPICS

Substituting for a Variable, Linear Equations, Computation, Conversion, Reading Data from a Table, Graphing, Problem Solving, Spreadsheets

TYPES OF INTELLIGENCES

Logical/Mathematical, Verbal/Linguistic, Intrapersonal, Interpersonal

CONCEPTS

Students will do the following:

1. Use a formula to calculate fluid requirements for dehydrated dogs.
2. Substitute for variables.
3. Graph weight/fluid ratio to show a linear relationship.
4. Using a computer, create a spreadsheet to perform repeated calculations (optional).

MATERIALS

- Copy of Worksheet for each student
- Copy of Graphing Sheet for each student

- Copy of Grading Matrix for each student
- Calculators
- Computers with spreadsheet software.

WHAT TO DO

Survey students to find out how many of them have dogs as pets. Ask how much their dogs weigh (approximately). Give each student a copy of the activity sheet and explain that they will be determining the rate of fluid per minute to set up an intravenous (IV) drip for a dog. Review the formula (on the activity sheet) and explain that it is used to calculate the amount of liquid a dog needs during one 24-hour period (ml/24 hr) but that they will need to convert the amount of liquid needed to milliliters per minute (ml/min).

Use the average weight of different breeds to demonstrate how to calculate the rate of fluid per minute. For example, if a dog weighed 6 pounds, substituting into the formula would result in 30 ml/1 lb × 6 = 180 ml/day. To convert ml/day to ml/min, students must use conversion strategies; for example, 1 day = 1,440 minutes, therefore, 180 ml/day 180 ml/1,440, or 0.125.

After students have calculated the amount of fluid needed for each of the breeds listed on the activity sheet, they graph this data on the graphing sheet. Student graphs should reflect a linear function since the weights are multiplied by a constant and there is a positive correlation between the dog's weight and the amount of fluid needed.

VARIATION

Students can take a survey of breeds of dogs owned by their classmates. They can research the average weight of the dogs using the Internet or library resources (being careful to determine if the dog is a male or female) and calculate fluid requirements for all of the dogs owned by students in the class. Male dogs weigh more than female dogs and (for accuracy) the differentiation is interesting. If they have access to computers and spreadsheet software, they can record this data in a spreadsheet (see Tie to Technology section).

TIE TO TECHNOLOGY

Using a computer, students can create spreadsheets that perform repeat calculations. They can enter the data collected about their classmates' dogs and let the computer calculate how much fluid each dog needs per day or minute. A sample spreadsheet for different breeds is shown below. One table is shown as it typically appears, and the other table is shown with the cell formulas indicated. By changing the weights in column B, the amounts of fluid per 24 hours and per minute are changed automatically. Column D converts the drip to milliliters per minute. The formula in each column can be brought down using the "fill down" function. Formulas like these show students the power of using spreadsheets.

	A	B	C	D
1	**Breed of Dog**	**Average Weight**	**ml/24 hrs**	**ml/min**
2	Terrier	20.5	615	0.427
3	Malamute	85	2550	1.771
4	Chihuahua	8	240	0.167
5				

	A	B	C	D
1	**Breed of Dog**	**Average Weight**	**ml/24 hrs**	**ml/min**
2	Terrier	20.5	=B2*30	=C2/1440
3	Malamute	85	=B3*30	=C3/1440
4	Chihuahua	8	=B4*30	=C4/1440
5				

Note: The grading matrix includes a section for the use of a spreadsheet. If this is not part of the activity, the matrix will need to be changed.

ASSESSMENT

1. Student products
2. Grading matrix
3. If spreadsheets are used, assess the quality of the formulas and setup

ON THE INTERNET

The official website for the American Kennel Club (see appendix for URL) provides a plethora of information about breeds of dogs, their care, dog shows, and other related topics.

Vet Math

Worksheet

Do you own a dog? Do you have a friend that does? What type of dog is it? What is its weight? If the dog became dehydrated and needed intravenous fluids, how many milliliters of fluid would it need each minute? Use this formula to determine the amount of fluid a dog needs during a 24-hour period. Note that it must be converted to the milliliters needed per minute.

$$\frac{\textbf{30 ml}}{\textbf{1 lb}} \times \textbf{weight (lb) of animal = milliliters needed per 24 hours}$$

Show your work here (be sure to include the breed and weight of the dog).

Breed of Dog	**Weight in lb (Average wt of males)**	**ml/24 hr**	**ml/min**
Scottish Terrier	20 1/2		
Alaskan Malamute	85		
Golden Retriever	70		
Chihuahua	8		
Collie	67 1/2		
Bulldog	50		
Labrador Retriever	60		
Newfoundland	140		
Portuguese Water Dog	45		

Vet Math

Graphing Sheet

Directions

Use the following grid to graph the relationship between the weight of a dog and the milliliters of fluid needed per minute.

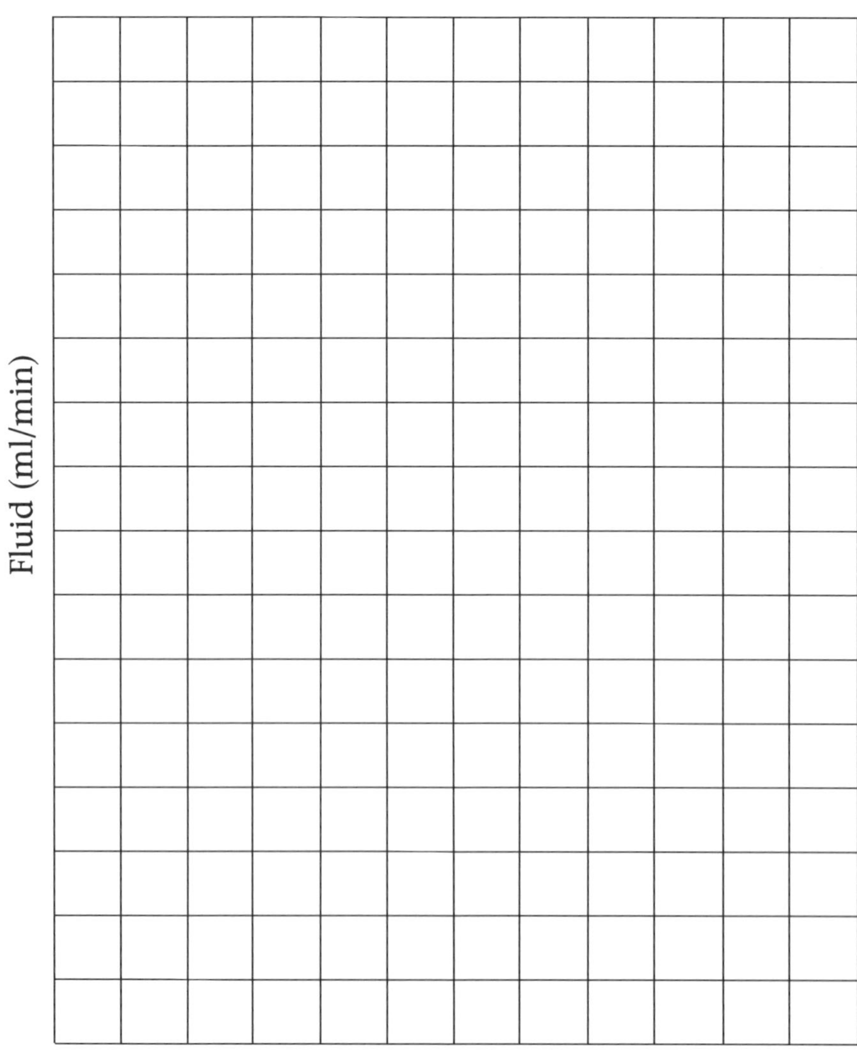

Vet Math

Grading Matrix

Names ____

Date____________________ Class ______________________

Criteria	4	3	2	1
Accuracy of computation				
Accuracy of graph				
Quality of spreadsheet				
Comments:				

4 = Superlative, 3 = Competent, 2 = Limited, 1 = Inadequate

Seeing to the Horizon: Formulas in the Real World

MATH TOPICS

Using Formulas, Technology and Spreadsheets, Computation, Square Roots, Reading Data from a Table, Averages, Problem Solving

TYPES OF INTELLIGENCES

Logical/Mathematical, Verbal/Linguistic, Interpersonal

CONCEPTS

Students will do the following:

1. Use data to calculate the distances one could see from the tops of ten of the tallest habitable buildings to the horizon.
2. Work in a group of four to estimate the height of their school building and to find the distance to the horizon from this point.
3. Design a spreadsheet program for their data and automatically calculate the distances.

MATERIALS

- Copy of Activity Sheet for each student
- Copy of Grading Matrix for each group

- Overhead transparency of Class Data Table
- Scientific calculators
- Rulers, yardsticks, and/or tape measures
- Computers and spreadsheet software

WHAT TO DO

Distribute a copy of Activity Sheet to each student and discuss the formula with students. Explain that 1.22 is a constant and is based upon physical principles relating to Earth and its environment. Using scientific calculators, students should calculate the distances they can see from the top of the buildings listed in the data table. The answers (rounded to the nearest mile) are shown in the following table.

10 of the Tallest Habitable Buildings

Building	Height (in ft)	Distance You Can See to the Horizon (in mi)
Sears Tower	1,454	47
World Trade Center	1,368	45
Empire State Building	1,250	43
Amoco Building	1,136	41
John Hancock Center	1,127	41
Chrysler Building	1,046	39
Texas Commerce Tower	1,002	39
Allied Bank Plaza	992	38
First Canadian Plaza	952	38
International Building	950	38

Source: WTB Magazine, <http://www.worldstallest.com/96/fansky.html>, August 2, 1999.

Next, ask students to estimate the height of their school building. Have them form groups of four, give them tape measures, yardsticks, etc., and let them problem solve the height of the school. These estimates should be recorded on the class data table (overhead transparency) and discussed: Do they make sense? How did students decide how tall the building is? What is the range of the estimates? Do some estimates appear to be outliers (far from the mean)? Before an average or mean is found, should some of the estimates be eliminated? Once all of the data is accepted and a class average is found, this estimate can be used to calculate the distance to the horizon from the top of the school building.

VARIATION

Using Internet or library resources, students can research the tallest mountains on Earth and calculate the distance to the horizon from the highest point.

TIE TO TECHNOLOGY

Sample spreadsheets showing the height and distance to the horizon for three of the buildings listed on the data table are shown below. In each spreadsheet, column B lists the heights of three buildings (i.e., Sears Tower, World Trade Center, and Empire State Building). In the first table, column C shows the distance to the horizon. In the second table, column C shows a formula students can use to calculate the distance to the horizon.

	A	B	C
1	Building	Height in ft.	Distance to Horizon
2	Sears Tower	1454	47
3	World Trade	1368	45
4	Empire State	1250	43
5			

	A	B	C
1	Building	Height in ft.	Distance to Horizon
2	Sears Tower	1454	=1.22*SQRT(B2)
3	World Trade	1368	=1.22*SQRT(B3)
4	Empire State	1250	=1.22*SQRT(B4)
5			

ASSESSMENT

1. Student products
2. Observation and questioning: "Explain how your group estimated the height of the school."

3. Journal question: "Explain the formula you used to find the distance from the estimated height of the school building to the horizon. Find the height of a structure (perhaps on the Internet) and use the formula to calculate how far you can see to the horizon from the top of that structure."

4. Grading matrix

ON THE INTERNET

Boston College has set up an extensive site, A Digital Archive of American Architecture (see appendix for URL), that offers a wealth of information on skyscrapers. It defines the term *skyscraper,* lists and shows pictures of the skyscrapers of the nineteenth and twentieth centuries, and has links to other architectural websites. As new buildings are built (bigger and better), you can access this site to update the information in this lesson.

Seeing to the Horizon: Formulas in the Real World

Activity Sheet

The distance you can see to the horizon on a clear day is given by the formula

$$d = 1.22\sqrt{h}$$

In this formula, d represents the distance in miles and h represents the height in feet your eyes are from the ground. In other words, the higher you are, the further you can see.

Directions

Use this formula to determine for each building the distance you could see to the horizon if you were standing on its roof.

10 of the Tallest Habitable Buildings		
Building	**Height (in ft)**	**Distance You Can See to the Horizon (in mi)**
Sears Tower	1,454	
World Trade Center	1,368	
Empire State Building	1,250	
Amoco Building	1,136	
John Hancock Center	1,127	
Chrysler Building	1,046	
Texas Commerce Tower	1,002	
Allied Bank Plaza	992	
First Canadian Plaza	952	
International Building	950	

Source: WTB Magazine, <http://www.worldstallest.com/96/fansky.html>, August 2, 1999.

Work with your group to estimate the height (in feet) of your school building.

Write your estimate here: ____

Write your estimate on the class data table.

Seeing to the Horizon: Formulas in the Real World

Class Data Table

Group	Estimate of Height of Building
Class Mean	

Be prepared to discuss the steps your group used to problem solve the height of the school building.

Seeing to the Horizon: Formulas in the Real World

Grading Matrix

Names ______________________________

Date__________________ Class ______________________________

Criteria	4	3	2	1
Accuracy of computation				
Quality of explanation of problem solving (estimation)				
How well did group estimate the height of the school building?				
How well did group work together?				
Comments:				

4 = Superlative, 3 = Competent, 2 = Limited, 1 = Inadequate

ACTIVITY 12

Falling Objects: Formulas in the Real World

MATH TOPICS

Substitution for a Variable, Using Formulas, Square Roots, Rounding, Problem Solving, Conversions, Spreadsheets

TYPES OF INTELLIGENCES

Logical/Mathematical, Bodily/Kinesthetic, Verbal/Linguistic, Interpersonal

CONCEPTS

Students will do the following:

1. Substitute for a variable in a formula.
2. Use formulas to find the time required for an object to fall.
3. Use square roots and round their answers.
4. Convert their data from feet/second to miles/hour.
5. Use a spreadsheet to compute repetitive calculations.
6. Conduct an experiment to determine how long it will take for a marshmallow to fall from the roof of their school building. Using the formula $d = 16t^2$, they can calculate the height of their school building.

MATERIALS

- Copy of Worksheet for each student
- Copy of Activity Sheet for each group
- Copy of Grading Matrix for each group
- Marshmallows
- Calculators
- Stopwatches
- Computers, spreadsheet software

WHAT TO DO

The time it takes for something to fall is determined by the formula used to find the vertical drop of an object. It assumes that there is no wind resistance and that the force of gravity is constant on every object (regardless of mass). On Earth, objects accelerate at a rate of 32 feet per second per second. This acceleration is the result of the force of gravity that pulls an object toward the center of Earth. This constant is used in the formula as a substitute for the variable *g,* but when the fraction is reduced, it becomes 16.

To find the time it takes to reach the ground, students substitute into the formula:

$$t = \sqrt{\frac{2d_v}{g}} = \sqrt{\frac{2d_v}{32}} = \sqrt{\frac{d_v}{16}}$$

where d_v = vertical distance and g = gravity

Following order of operations, students can multiply the height of the building by 2, then divide it by 32 (the acceleration caused by gravity), then find the square root. Or by simplifying the radical, students can divide the height of the building by 16 and then find the square root.

To find the rate of speed of the marshmallow, students must convert *ft/sec* to *mi/hr*. There are a number of ways of doing this; following is one way:

1. Find the rate of speed in *ft/sec* by multiplying the time (third column) by 32. This rate is *ft/sec.*

2. Convert *ft/sec* to *mi/hr*.

Following is the computation for the Sears Tower.

1. 9.5 sec × 32 ft/sec^2 = 304 *ft/sec*

2. 304 *ft/sec* × 3,600 *sec/hr* = 1,094,400 *ft/hr*

3. 1,094,400 *ft/hr* × 1 *mi/5,280 ft* ≈ 207 *mi/hr*

The answers to each of the problems are shown in the following table. The rate of speed has been rounded to the nearest mile per hour.

Ten of the Tallest Habitable Buildings

Building	Height (in ft)	Time (to the nearest 1/10 sec)	Rate of Speed (to the nearest mi/hr)
Sears Tower	1,454	9.5	207
World Trade Center	1,368	9.2	201
Empire State Building	1,250	8.8	192
Amoco Building	1,136	8.2	178
John Hancock Center	1,127	8.1	177
Chrysler Building	1,046	8.1	177
Texas Commerce Tower	1,002	7.9	172
Allied Bank Plaza	992	7.9	172
First Canadian Plaza	952	7.7	168
International Building	950	7.7	168

Source: WTB Magazine, <http://www.worldstallest.com/96/fansky.html>, August 2, 1999.

Once students understand how to use the formulas, they can conduct a real-life experiment. To determine the height of their school building, they can use a variation of the vertical drop formula of $d = 16t^2$. Have a student (or another teacher) drop 5 marshmallows from the roof of the school one at a time. Students (in groups) on the ground can time the fall to the nearest tenth of a second. A class average can be found, and students can calculate the height of the building using the distance formula.

VARIATION

Have students research the highest roller coasters in the world. An interesting website for this data is World of Coasters (see appendix for URL). Using a computer, students can design a data collection table in a spreadsheet program for inputting formulas for the vertical drop and rate of speed. After typing in the heights of the roller coasters, the time and speed will automatically be calculated.

TIE TO TECHNOLOGY

When formulas such as these are used, the computation is often very repetitive and thus it is the perfect time to show students the power of spreadsheets in solving these types of problems. The following spreadsheet shows the formulas used for vertical drop (third column—C) and rate of speed when the object hits the ground (fourth column—D). The spreadsheet is displayed with formulas as well as a data table.

	A	B	C	D
1	**Building**	**Height in ft.**	**Time (sec)**	**Speed (mph)**
2	Sears Tower	1454	9.5	208
3	World Trade	1368	9.2	202
4	Empire State	1250	8.8	193
5				

	A	B	C	D
1	**Building**	**Height in ft.**	**Time (sec)**	**Speed (mph)**
2	Sears Tower	1454	=SQRT(B2/16)	=(C2*32*3600)/5280
3	World Trade	1368	=SQRT(B3/16)	=(C3*32*3600)/5280
4	Empire State	1250	=SQRT(B4/16)	=(C4*32*3600)/5280
5				

ASSESSMENT

1. Student products (activity sheets)
2. Observation and questioning: "Explain how you converted feet/second to miles/hour" or "How close were the times you got for the marshmallow drop? Was there a difference in times? If there was, why do you think this occurred?"
3. Journal question: "Why did we drop 5 marshmallows from the roof and use the mean, or average, for our calculations?"
4. Grading matrix

Falling Objects: Formulas in the Real World

Worksheet

When something falls off of a building, it is called a free-falling object. To figure out how long it will take to hit the ground, it's necessary to know the distance the object fell. On Earth, objects accelerate at a rate of 32 ft/sec^2. The formula that is used to calculate how long it will take to fall is

$$t = \sqrt{\frac{2d_v}{g}} \text{ or } t = \sqrt{\frac{2d_v}{32}} \text{ or } t = \sqrt{\frac{d_v}{16}}$$

In the formula, t is the time the object falls, d_v is the vertical drop, and g is the acceleration due to gravity. By simplifying the formula, we find that the time it takes for an object to fall is equal to the square root of the distance divided by 16.

Directions

If someone stood on the top of one of the ten tallest habitable buildings and dropped a marshmallow, how long would it be before it reached the ground? How fast would it be traveling when it hit the ground? Use the vertical drop formula and the data in the following table to find this information for each of the listed buildings.

Ten of the Tallest Habitable Buildings

Building	Height (in ft)	Time (to the nearest 1/10 sec)	Rate of Speed (32 ft/sec/sec)
Sears Tower	1,454		
World Trade Center	1,368		
Empire State Building	1,250		
Amoco Building	1,136		
John Hancock Center	1,127		
Chrysler Building	1,046		
Texas Commerce Tower	1,002		
Allied Bank Plaza	992		
First Canadian Plaza	952		
International Building	950		

Source: WTB Magazine, <http://www.worldstallest.com/96/fansky.html>, August 2, 1999.

Falling Objects: Formulas in the Real World

Activity Sheet

Now let's conduct a little experiment. How tall do you think your school building is? Discuss this with your group and write your estimate here: ____

Directions

1. Five marshmallows will be dropped from the roof of your school. Time their fall as carefully as you can to the nearest 1/10 of a second.
2. Record the 5 trials on this activity sheet.
3. Find the average.
4. Use a variation of the vertical drop formula ($d = 16t^2$) to calculate the distance the marshmallows traveled. Use your group average for *t*.

$$t = \sqrt{\frac{d}{16}}$$

Trial	Time To the nearest 1/10 sec
1	
2	
3	
4	
5	
Average	

Based upon our data, we calculate the height of the school to be ________

Our work:

Falling Objects: Formulas in the Real World

Grading Matrix

Names ______________________________

Date______________ Class ______________________

Criteria	4	3	2	1
How accurate was the use of formulas?				
Accuracy of measurement conversions				
Quality of data collection				
How well did group apply the formulas to the experiment?				
How well did group work together?				
Comments:				

4 = Superlative, 3 = Competent, 2 = Limited, 1 = Inadequate

The 18-Hour Clock: Modular Designs

MATH TOPICS

Finite Mathematics, Computation, Problem Solving, Linear Equations, Substitution for a Variable

TYPES OF INTELLIGENCES

Visual/Spatial, Logical/Mathematical, Bodily/Kinesthetic

CONCEPTS

Students will do the following:

1. Develop a linear equation
2. Design and complete a table for mod 18
3. Compute in modular (clock) arithmetic
4. Develop a design using mod 18 numbers

MATERIALS

- Copy of Circular Design Sheet for each student
- Copy of Grading Matrix for each student
- Overhead transparency of mod 12 circle with connecting lines
- Ruler, compass, and protractor for each student
- Colored pencils or fine-tip markers

WHAT TO DO

Ask students to look at the clock in the classroom and imagine that it is 10 o'clock in the morning. They have to catch a plane in 5 hours. At what time will they be getting on the plane? (Inform them that you are not interested in military or universal time.) The answer is 3 o'clock in the afternoon. What students have done is add $10_{\text{mod } 12} + 5_{\text{mod } 12} = 3_{\text{mod } 12}$

Looking at a clock (the 12 is represented by a 0), imagine starting out at 10:00 and looking 5 hours ahead. You end up at 3:00—so reading a clock is the same as adding in mod 12.

Work with students to develop a table for mod 12 using this equation: y = x + 7. The finished table will look like this:

x	0	1	2	3	4	5	6	7	8	9	10	11
y	7	8	9	10	11	0	1	2	3	4	5	6

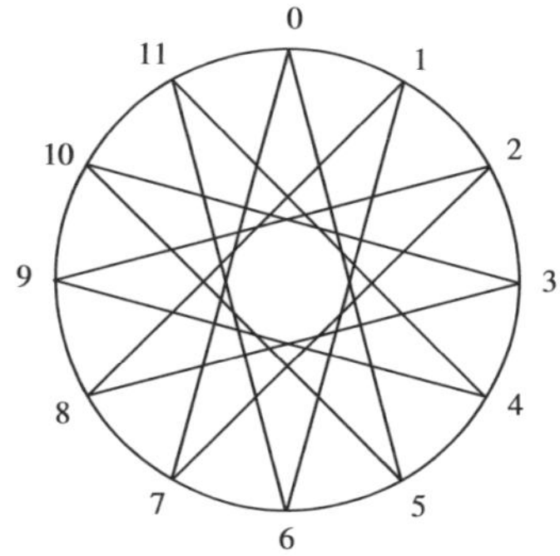

Show an overhead transparency of a mod 12 circle with connecting lines.

Have students draw a circle using their compasses and divide their circle into 12 equal parts—each angle will be 30°. The arc of the circle must be numbered as shown in the overhead example. Explain that they are to connect the corresponding *x* and *y* values as indicated on the table. In other words, 0 connects to 7, 1 connects to 8, and so on.

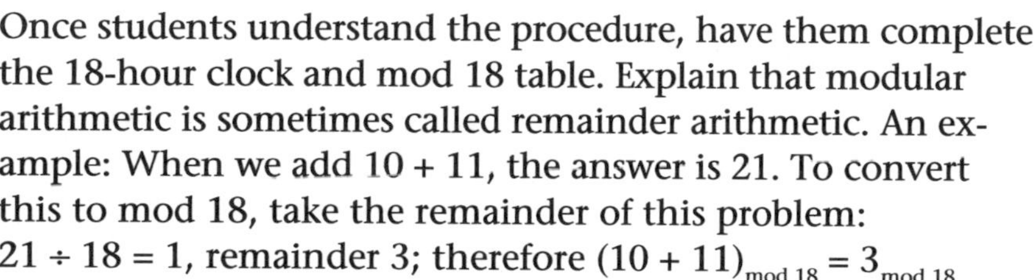

Once students understand the procedure, have them complete the 18-hour clock and mod 18 table. Explain that modular arithmetic is sometimes called remainder arithmetic. An example: When we add 10 + 11, the answer is 21. To convert this to mod 18, take the remainder of this problem: $21 \div 18 = 1$, remainder 3; therefore $(10 + 11)_{\text{mod } 18} = 3_{\text{mod } 18}$

If the same equation (y = x + 7) were graphed, the table would look like this:

x	0	1	2	3	4	5	6	7	8	9	10	11	12	13	14	15	16	17
y	7	8	9	10	11	12	13	14	15	16	17	0	1	2	3	4	5	6

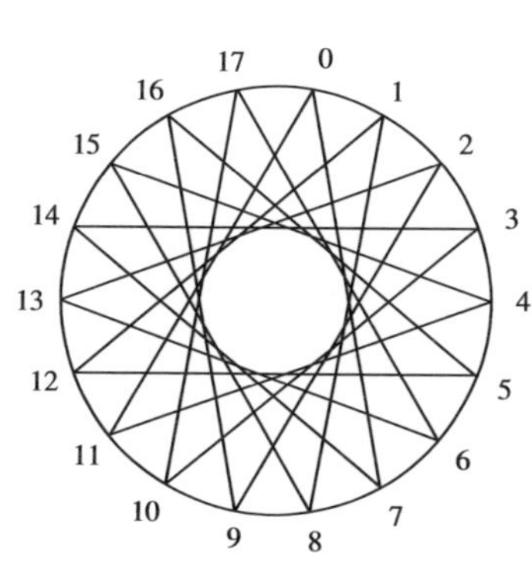

and the design would look like the one on the left.

Students can devise their own equations and connect the *x* values to the *y* values on the circle. By creatively coloring in the designs, students can see that all of their equations have at least one line of symmetry. If students create an equation in which the *x* and *y* values are too similar, tell them to make up a new one because they will be very disappointed in the design.

VARIATION

Students can draw circles in mod 36 (circles with labels of 0 through 35). Each angle will be 10° and the designs will be much more intricate.

ASSESSMENT

1. Student products
2. Observation and questioning: "Explain how you developed the coordinates on the arc of the circle" or "Have you found one or more lines of symmetry in your circle design?"
3. Journal question: "Explain why modular arithmetic is sometimes called clock arithmetic. Why is it appropriate that the model used is a circle rather than a number line?"
4. Grading matrix

ON THE INTERNET

The website Symmetry and Pattern: The Art of the Oriental is a collaborative project of The Math Forum and The Textile Museum (see appendix for URL). It is devoted to the relationship between geometric design and art and contains activities related to symmetry and patterns, explores transformations (translations, reflections, glide reflections, and rotations), and has an extensive page for educational resources. Other sites can be found by using the phrases "math and art" or "math and symmetry."

The 18-Hour Clock: Modular Designs

Circular Design Sheet

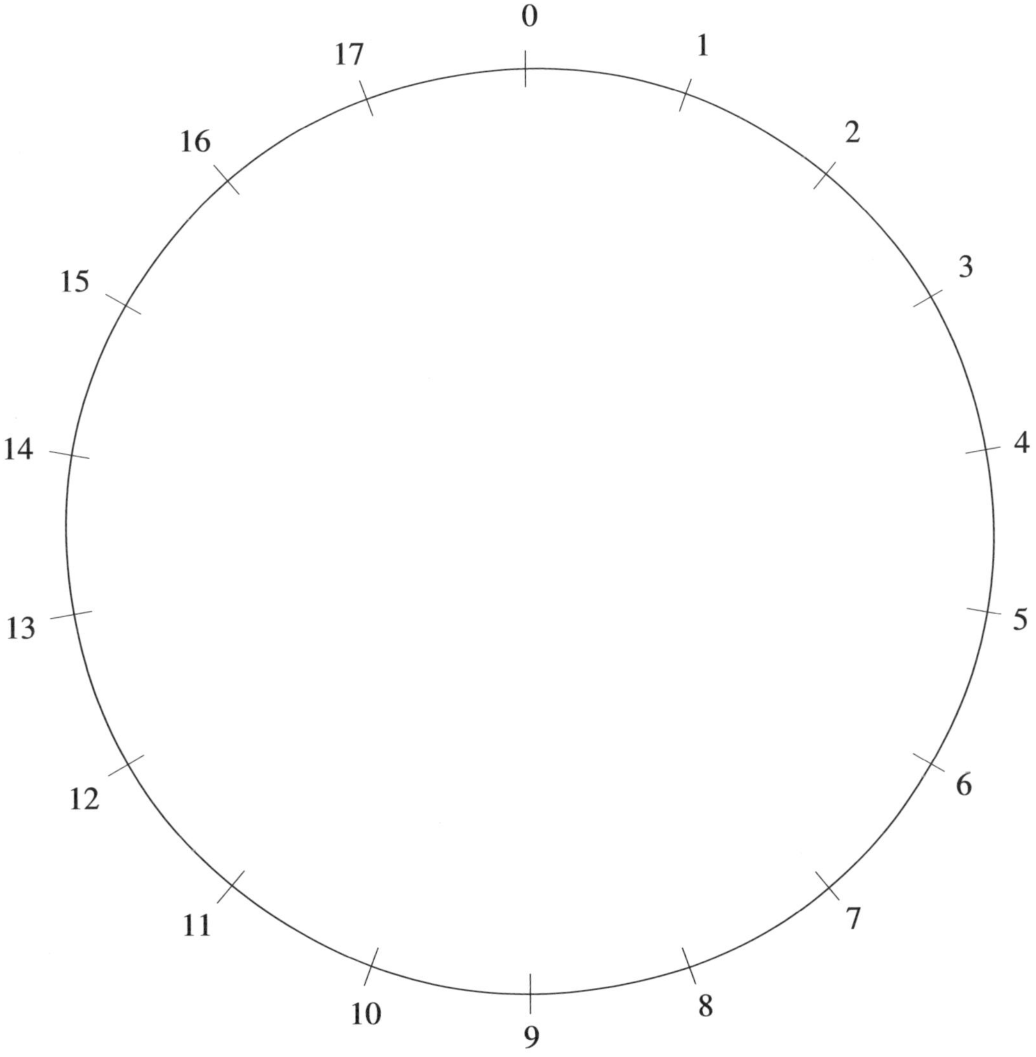

0 1 2 3 4 5 6 7 8 9 10 11 12 13 14 15 16 17

x	0	1	2	3	4	5	6	7	8	9	10	11	12	13	14	15	16	17
y																		

The 18-Hour Clock: Modular Designs

Grading Matrix

Names ____

Date____________________ Class ________________________

Criteria	4	3	2	1
Accuracy of modular arithmetic table				
Accuracy of modular design				
Originality of linear equation				
Overall quality of project				

Comments:

4 = Superlative, 3 = Competent, 2 = Limited, 1 = Inadequate

CHAPTER 3

Measurement and Geometry

CHAPTER 3

Measurement and Geometry

The knowledge at which geometry aims is the knowledge of the eternal.—Plato

Measurement and geometry are the strands of mathematics that help people describe the physical world in which they live. The study of geometry helps students see the world of mathematics in a different light—the mathematics of shapes and forms, not numbers and formulas.

The study of measurement requires students to directly interact with their environment. When students learn to measure, they can answer such questions as "How much will that hold?" or "How tall is it?" They learn about weight, distance, and time. They develop *measurement sense* as they acquire a true understanding of space and size.

It is important that students be actively involved in their learning to fully understand the concepts. The experiences in this chapter encourage students to work collaboratively and develop their multiple intelligences while they learn more about their physical world.

Does the height from which a ball is dropped affect how high it bounces? Students answer this intriguing question by conducting the experiment outlined in the activity The Bouncing Ball. Working collaboratively, students collect, organize, analyze, and graph their data and, in the process, learn a variety of measurement skills and concepts.

Soda Pop Math combines the study of two- and three-dimensional geometry. Beginning with a sheet of posterboard, students design and carefully measure a cylindrical net. By working collaboratively to design a real product, students use a variety of intelligences while learning about the attri-butes of a cylinder.

Is it easier to hit the outside of the dartboard than the center? Students are invited to synthesize the strands of measurement, geometry, and probability to answer this question. In The Dartboard: What Are The Chances?, students explore the areas of concentric circles and then employ their new skills to design their own game. By assigning points (based upon the geometric probability of each event), they assimilate and refine their knowledge.

Students explore the connections between the physics of musical frequencies and the patterns of mathematics in Math and Music: Frequencies. With their newfound knowledge, they build a musical instrument and write original music.

Paper-folding is a remarkable and informal way to develop an understanding of geometric concepts. Students can feel and experience the shapes and sizes of different geometric shapes as they fold and manipulate paper into polygons. The activity Paper-Folding Polygons permits students to experience a regular pentagon and use concepts of symmetry to transform the shape into a not-so-ordinary snowflake!

The Valley of Mars is a delicious way to use nonstandard units of measures (i.e., Oreo cookies) to find the length of a valley that is thirteen times greater than that of the Grand Canyon. After groups discover "how many Oreo cookies long" the canyon is, they calculate how expensive a measuring tool these cookies really are.

No student really believes he or she will ever have to know how to find the volume of a sphere. When this topic appears in the mathematics text, the first question typically is "When are we ever going to use this?" But in the activity Limitless Lung Power, students learn how much hot air they have in their lungs by calculating the volume of a balloon. By blending the mathematical strands of data collection, solid geometry, and statistics, students are able to make real-world connections.

G. H. Hardy once described mathematics as "having a very high degree of unexpectedness, combined with inevitability and economy" (Locher 1971, 2). These words can be used to describe the artwork of M. C. Escher. As Escher said:

> By keenly confronting the enigmas that surround us, and by analyzing the observation that I had made, I ended up in the domain of mathematics. Although I am absolutely innocent of training or knowledge in the exact sciences, I often seem to have more in common with mathematicians than with my fellow artists" (1986, 2–3).

Escher was fascinated with the art of Islam and studied Moorish tilings in Spain. His work has become very popular and, thus, the activity Tessellations: Transformation a la Escher will have built-in fascination for students. At the same time, they will be learning transformational geometry—the geometry of movement within a plane. This activity introduces students to one of the transformations—translations, or slides. By using a simple cut-and-paste, step-by-step approach, this activity almost guarantees every student will experience success. Those students with a highly developed visual/spatial intelligence will think they made a wrong turn and ended up in their favorite class—art!

Enjoy these motivating and enriching activities and projects with your students. Each allows students to touch, encounter, and experience the mathematics of geometry and measurement.

ACTIVITY 14

The Bouncing Ball

MATH TOPICS

Data Collection, Measurement, Averages, Graphing

TYPES OF INTELLIGENCES

Logical/Mathematical, Bodily/Kinesthetic, Interpersonal, Verbal/Linguistic, Intrapersonal

CONCEPTS

Students will do the following:

1. Conduct an experiment to observe how high a ball bounces when dropped from four different heights.
2. Record and average their data in a table.
3. Graph their results and find a "line of best fit."

MATERIALS

- Overhead transparencies of Data Collection Sheet and Graphing Sheet
- Copy of Data Collection Sheet for each group
- Copy of Graphing Sheet for each group
- Copy of Experimental Analysis Sheet for each student

- Copy of Grading Matrix for each group
- Meter stick and masking tape for each group
- Rubber ball for each group
- Calculators

WHAT TO DO

Have students form into groups of four. Ask students if they think a ball will bounce higher if it is dropped from a greater distance. Explain that in this experiment, a ball will be dropped from heights of 100 cm, 80 cm, 60 cm, and 40 cm. Ask students whether they think the heights the ball bounces will differ based on the distance from which it was dropped. Have students discuss the possibilities in their groups.

Each group will need a meter stick and masking tape. The sticks must be attached to a table at a 90° angle to the floor. The students will perform and rotate the following tasks: 1) dropping the ball from the specified height, 2) keeping track of how high the ball bounces back during the three trials, 3) recording the data, and 4) finding the average of the three trials.

Each of the trial heights should be recorded on the Data Collection Sheet. Students transfer this information to the Graphing Sheet and draw a line to represent the trend of the data.

Once the data is graphed, ask students to write a verbal description of the experiment along with their analysis of the results on the Experimental Analysis Sheet. A copy of this sheet is given to each group member and the written analysis done as an individual activity.

VARIATION

The experiment can be extended to include different types of balls—the ball types become the experimental variable. The height remains constant and the type of ball varies. There should be three trials for each type of ball and an average determined. Students can be required to designate the dependent and independent variables in this revised experiment.

TIE TO TECHNOLOGY

Students can conduct another experiment (of the same type) using different types of balls but dropping them from the same height. They can then use a computer or graphing calculator to record and display results. Data can be collected and entered into a graphing calculator table or a computer spreadsheet. Students can generate multiple line graphs or bar graphs to analyze the data and compare the actions of each of the balls.

Another technology extension involves the use of graphing calculators. Some graphing calculators can be attached to Calculator Based Lab (CBL) stations. One of the labs available is a motion detector, which produces graphs based upon the intensity of the motion—how close the movement is to the sensor. This lab would fit well with this experiment.

ASSESSMENT

1. Student products
2. Observation and questioning: "Who in the group can explain why you made the predictions you did? Do your predictions appear to be correct?"
3. Journal question: "Research the dependent and independent variables used in this experiment and describe how they affected the results."
4. Grading matrix

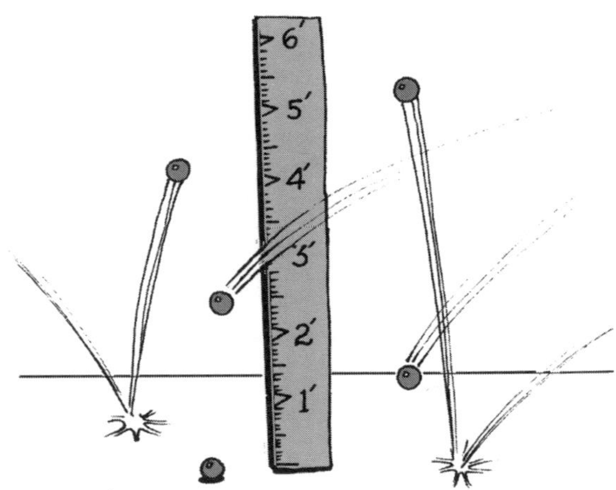

The Bouncing Ball

Data Collection Sheet

Directions

Tape a meter stick to the side of a table as shown at the right. Be sure that the stick is perpendicular to the floor. Drop the ball from each of the heights indicated on the table three times and find the average height the ball bounces back on the first bounce. Measure the height, as carefully as you can, from the bottom of the ball. Record your results in the following tables. After your group has completed the experiment, graph your results on the Graphing Sheet.

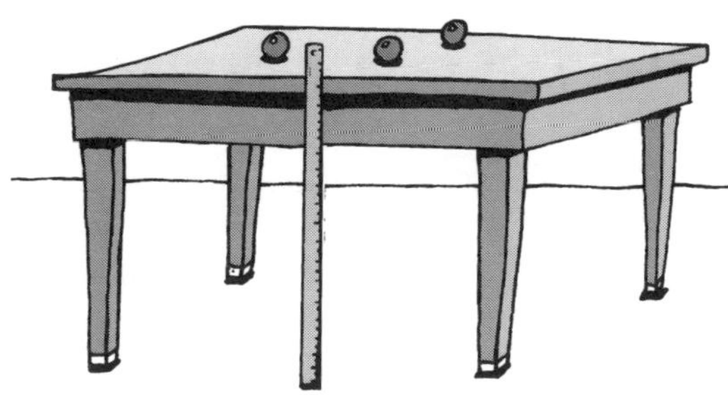

Height Ball Dropped 100 cm Trials	Height Ball Bounced (in cm)	Average of Three Trials
Trial 1		
Trial 2		
Trial 3		
Height Ball Dropped 80 cm Trials	Height Ball Bounced (in cm)	Average of Three Trials
Trial 1		
Trial 2		
Trial 3		
Height Ball Dropped 60 cm Trials	Height Ball Bounced (in cm)	Average of Three Trials
Trial 1		
Trial 2		
Trial 3		
Height Ball Dropped 40 cm Trials	Height Ball Bounced (in cm)	Average of Three Trials
Trial 1		
Trial 2		
Trial 3		

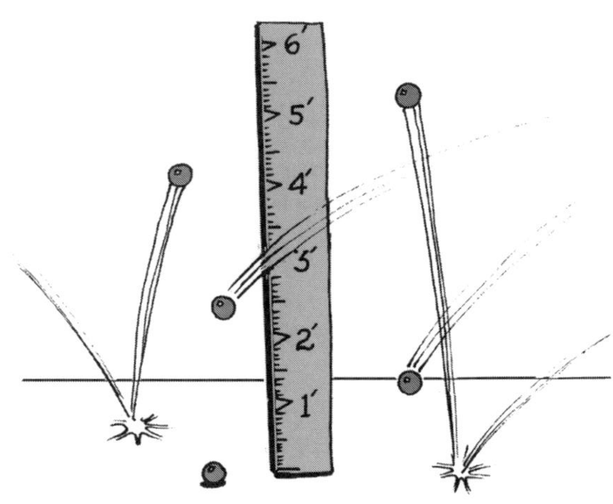

The Bouncing Ball

Graphing Sheet

Directions

Label each axis using equal units. Then locate and graph your data. Do not connect the points but draw a "line of best fit," or a line that best illustrates the trend of the data. Describe the trend of the data on the sheet provided.

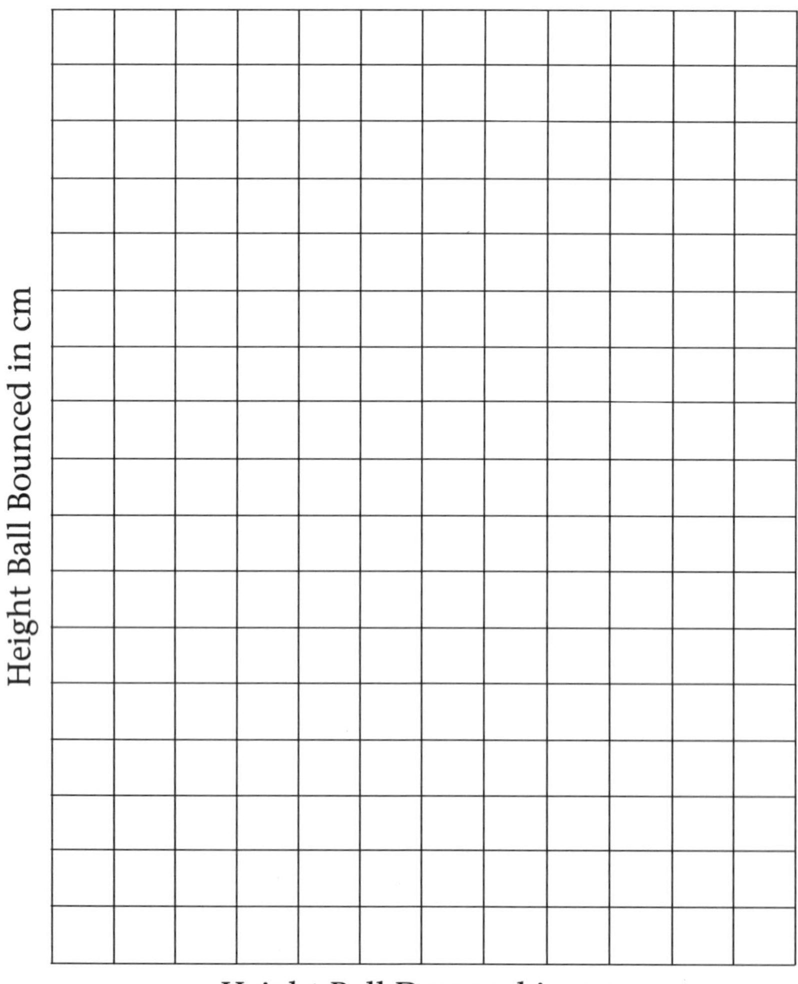

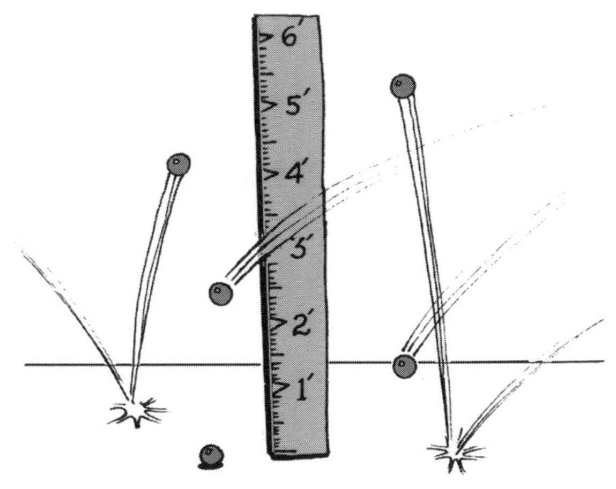

The Bouncing Ball

Experimental Analysis Sheet

Describe the results of your experiment. Be sure to clearly define the variables and analyze the effect of height on the results.

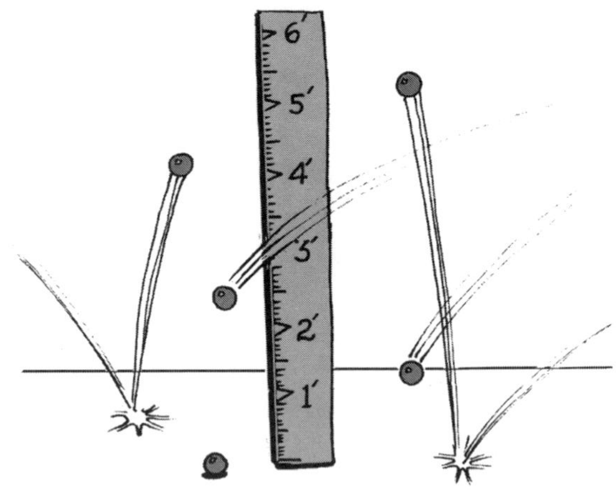

The Bouncing Ball

Grading Matrix

Names ____

Date____________________ Class ______________________________

Criteria	4	3	2	1
Quality of data collection				
Quality of graph				
Quality of written analysis				
How well did group work together?				
Comments:				

4 = Superlative, 3 = Competent, 2 = Limited, 1 = Inadequate

ACTIVITY 15

Soda Pop Math

MATH TOPICS

Geometry, Spatial Reasoning, Problem Solving, Measurement

TYPES OF INTELLIGENCES

Logical/Mathematical, Visual/Spatial, Verbal/Linguistic, Interpersonal, Bodily/Kinesthetic

CONCEPTS

Students will do the following:

1. Work collaboratively to problem solve transforming a flat surface to a three-dimensional polyhedron.
2. Design a cylinder.
3. Design the surface of a cylinder (product can) to conform with consumer and nutritional issues.

MATERIALS

- Copy of Activity Sheet 1 for each student
- Copy of Activity Sheet 2 for each group
- Copy of Grading Matrix for each group

- Piece of posterboard (36 cm × 56 cm) for each group
- Compasses, rulers, and meter sticks
- Construction paper, glue, and scissors
- Markers, crayons, and colored pencils
- Calculators

WHAT TO DO

Discuss with students the idea of a net and how a model of a cube can be made from the net of six squares. Discuss the geometric shapes that make up a cylinder. Ask, "What is the relationship between the sides of a cylinder [the rectangle] and the top and bottom of a cylinder [the circles]?" Work through the discussion that precedes the directions on Activity Sheet 1. A net of a cylinder would look like this:

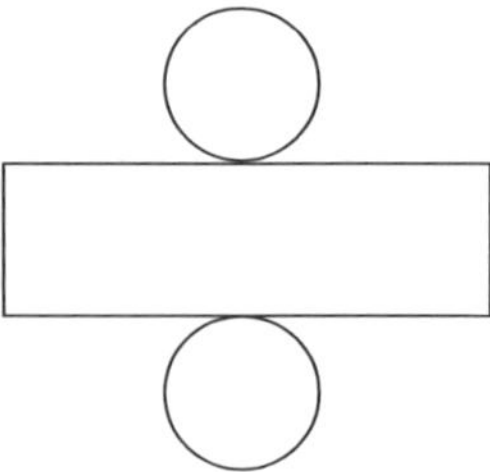

The length of the longer side of the rectangle is equal to the circumference of the cylinder and the length of the shorter side (width) of the rectangle is equal to the diameter of the circle.

Precise measurement is very important—stress to students that they cannot have gaps or overlapping edges.

VARIATION

Students can design nets for other three-dimensional shapes. An interesting project is designing nets for the five platonic solids—tetrahedron, hexahedron, octahedron, dodecahedron, and icosahedron.

TIE TO TECHNOLOGY

Many commercially available software programs encourage the design and analysis of two- and three-dimensional shapes. These programs, known as dynamic geometry software, permit students to "draw" polygons and polyhedrons, "measure" the sides and angles, "find the sum" of the angles, and calculate area and perimeter. Students also can resize the geometric shapes and evaluate the resulting changes.

Many of the more powerful programs allow students to use inductive reasoning to make suppositions based upon multiple trials with different shapes; for example, students can use the program to draw a variety of quadrilaterals, find the midpoints of each of the sides, and connect these midpoints. By analyzing the polygon produced in each of the cases, students can induce that the polygon created by connecting the midpoints of the sides of any quadrilateral is a parallelogram.

ASSESSMENT

1. Student products
2. Observation of student groups
3. Journal question: "Explain how an understanding of what geometric shapes a cylinder is made up of would help you find the surface area and volume of this three-dimensional shape."
4. Grading matrix

Soda Pop Math

Activity Sheet 1

- **A pop can (cylinder) is comprised of two basic geometric shapes. Can you name them?**
- **If you took off the top and bottom of a can, cut it down the sides, and opened it up, what polygon would you have?**
- **What would you call the length of this polygon (in relationship to the pop can)?**
- **What would you call the width of this polygon (also in relationship to the pop can)?**

Directions

Work in a small group to design a cylinder can.

1. Construct a can from a sheet of posterboard that is 36 cm × 56 cm, or 2,016 sq cm, without cutting out individual pieces. The flat pattern is called a *net.* Look at this net:

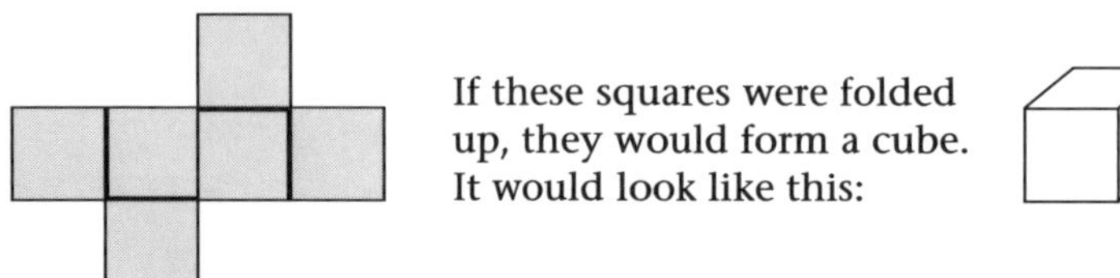

What do you think a net of a cylinder would look like?

Your job is to design a net that when folded will become a cylinder. Be sure to use as much of the posterboard as possible since any unused posterboard will be considered a waste of raw materials.

2. As a group, decide what product your can will hold and plan the can's design accordingly.
3. Decide what to show on the outside of the can. Most product cans show, at a minimum, the following:
   - Name of the product.
   - Advertising copy to help sell the product.
   - Attractive design to catch the buyer's eye.
   - Amount of product contained in the can (either by weight or volume).
   - Ingredients and nutritional information.

   Additionally, product cans may have
   - Recipe to give the buyer more of a reason to purchase it.
   - Directions on how to prepare contents.
4. Place your designs and information on the outside of the can prior to its construction. It will be much easier to do your work on a flat surface than on a curved one.

Soda Pop Math

Activity Sheet 2

Work with your group to complete this sheet. Analyze the size and shape of your cylindrical can. Measure to the nearest 1/10 cm.

1. How big was the piece of posterboard you used? __________
2. What are the dimensions of your design?
   a. Height _____
   b. Diameter of top _____
   c. Radius of top ______
3. What is your can's surface area? (This information would be necessary to determine the cost to design its surface.) _____
4. What is its volume? (To inform the buying public of the amount of product your can holds, you need to know its volume.) _____
5. If it cost $0.003 per sq cm to decorate the outside of the can (in order to attract the buying public), how much would you have to spend on this decoration? Explain how you got your answer.

Use the space below to do all of your computations.

Soda Pop Math

Grading Matrix

Names ______________________________

Date__________ Class ______________________

Criteria	4	3	2	1
Precision of cylindrical net				
Design of product can				
Overall quality of cylinder				
Accuracy of measurement and calculation on data sheet				
How well did group work together?				
Comments:				

4 = Superlative, 3 = Competent, 2 = Limited, 1 = Inadequate

ACTIVITY 16

The Dartboard: What Are the Chances?

MATH TOPICS

Probability Measurement, Using Formulas, Problem Solving

TYPES OF INTELLIGENCES

Logical/Mathematical, Visual/Spatial, Bodily/Kinesthetic, Interpersonal, Verbal/Linguistic

CONCEPTS

Students will do the following:

1. Measure the area of a series of concentric circles.
2. Find the ratio of the sizes of these circles.
3. Assign a point value to each circle.
4. Design a game using a point system and write the rules for it.

MATERIALS

- Copy of Activity Sheet for each group
- Copy of Game Sheet for each group
- Copy of Grading Matrix for each group
- Metric rulers
- Calculators (necessary because of the complexity of the computations)

WHAT TO DO

Discuss the concept of geometric probability—how the specific area of a dartboard can be equated with the number of points, or difficulty of landing, in that area. Ask students to observe the concentric circles of the dartboard and to estimate which circle has the largest area. Then, have students measure the diameter of each circle and calculate the area of each circle. (**Note:** The diameters of each of the concentric circles, starting with the smallest, are 1.5 cm, 3.5 cm, 5.5 cm, 7.5 cm, and 9.5 cm.)

Students should use the visible area of each of the circles to assign points in their games. They can use a variety of strategies to design their games, but the values should be based upon a ratio determined by the relative sizes of each of the circular areas. For example, the probability of the dart landing in the outermost circle (circle 1) is the ratio of the area of that circle to the total area (26.69/70.85 = 38%). It is important that this activity be as open-ended as possible and that students be given the opportunity to develop their own rules. For this reason, no attempt has been made to indicate what might be a possible solution.

VARIATION

Students can be asked to design another game board based upon geometric probabilities. These games can be drawn on posterboard and displayed at a grade-level "game fair."

TIE TO TECHNOLOGY

Following is an example of a spreadsheet that would permit students to find the areas and probabilities for concentric circles. Both numbers and possible formulas are shown.

	A	B	C	D	E
1	AREA OF CONCENTRIC CIRCLES				
2					
3	Diameter of 1	Diameter of 2	Diameter of 3	Diameter of 4	Diameter of 5
4	9.5	7.5	5.5	3.5	1.5
5	Radius of 1	Radius of 2	Radius of 3	Radius of 4	Radius of 5
6	4.75	3.75	2.75	1.75	0.75
7	Area of 1	Area of 2	Area of 3	Area of 4	Area of 5
8	26.69	20.41	14.13	7.85	1.76625
9	PROBABILITY-1	PROBABILITY-2	PROBABILITY-3	PROBABILITY-4	PROBABILITY-5
10	37.67%	28.81%	19.94%	11.08%	2.49%
11	TOTAL AREA				
12	70.84625				
13	TOTAL PROB.				
14	1				
15					

	A	B	C	D	E
1	**AREA OF CONCENTRIC CIRCLES**				
2					
3	**Diameter of 1**	**Diameter of 2**	**Diameter of 3**	**Diameter of 4**	**Diameter of 5**
4	9.5	7.5	5.5	3.5	1.5
5	**Radius of 1**	**Radius of 2**	**Radius of 3**	**Radius of 4**	**Radius of 5**
6	=A4/2	=B4/2	=C4/2	=D4/2	=E4/2
7	**Area of 1**	**Area of 2**	**Area of 3**	**Area of 4**	**Area of 5**
8	=(A6^2*3.14)-(B6^2*3.14)	=(B6^2*3.14)-(C6^2*3.14)	=(C6^2*3.14)-D6^2*3.14	=(D6^2*3.14)-(E6^2*3.14)	=E6^2*3.14
9	**PROBABILITY-1**	**PROBABILITY-2**	**PROBABILITY-3**	**PROBABILITY-4**	**PROBABILITY-5**
10	=A8/$A$12	=B8/$A$12	=C8/$A$12	=D8/$A$12	=E8/$A$12
11	**TOTAL AREA**				
12	=A8+B8+C8+D8+E8				
13	**TOTAL PROB.**				
14	=SUM(A10:E10)				

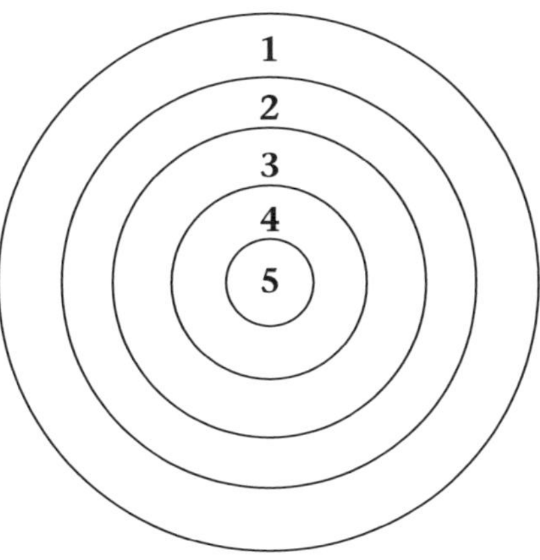

ASSESSMENT

1. Observation and questioning: "Can you explain how you developed the number of points you assigned to each of the regions of the dartboard?"

2. Journal question: "You are given a game board that looks like this one."

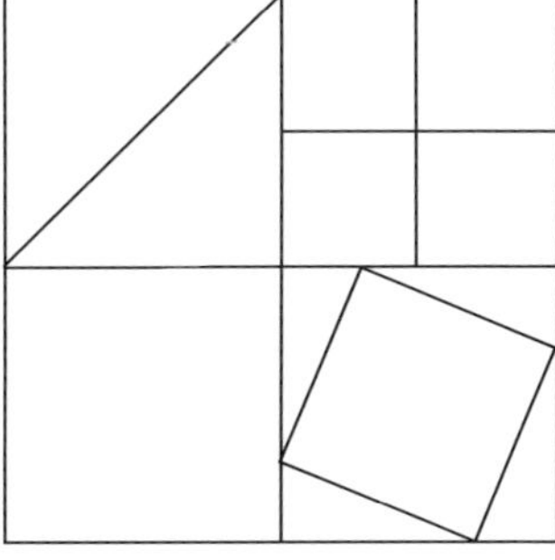

"Based upon the area of each of the regions, assign points. Explain your answer."

3. Grading matrix

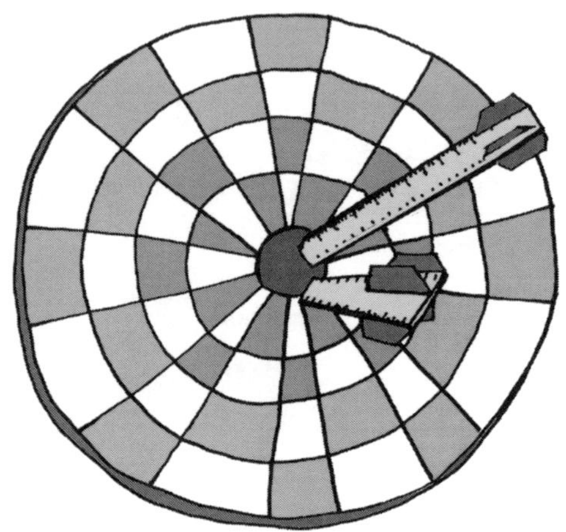

The Dartboard: What Are the Chances?

Activity Sheet

Have you ever wondered how points are assigned to a dartboard? Why are some spaces worth more than others? Are some areas easier to land on? Does it make sense that larger areas would be easier to hit? Let's investigate these questions by using what we know about the area of circles.

Directions

Work in small groups.

1. Measure each of the concentric circles in the dartboard shown below. These circles are concentric because they have the same center but different radii.
2. Find the area of the center circle. Use 3.14 for π.
3. Find the area of the next largest circle by finding the area of the entire circle and then subtracting the area of the circle within it (the smaller circle).
4. Continue finding the area of each of the circles in the same manner.
5. Assign points, or a point value, to each circle. Remember, the larger the area, the easier it should be to land a dart within it.
6. Design a game using the point system you devised. Be sure to explain the rules of your game and how you calculated the value of each of the circles.

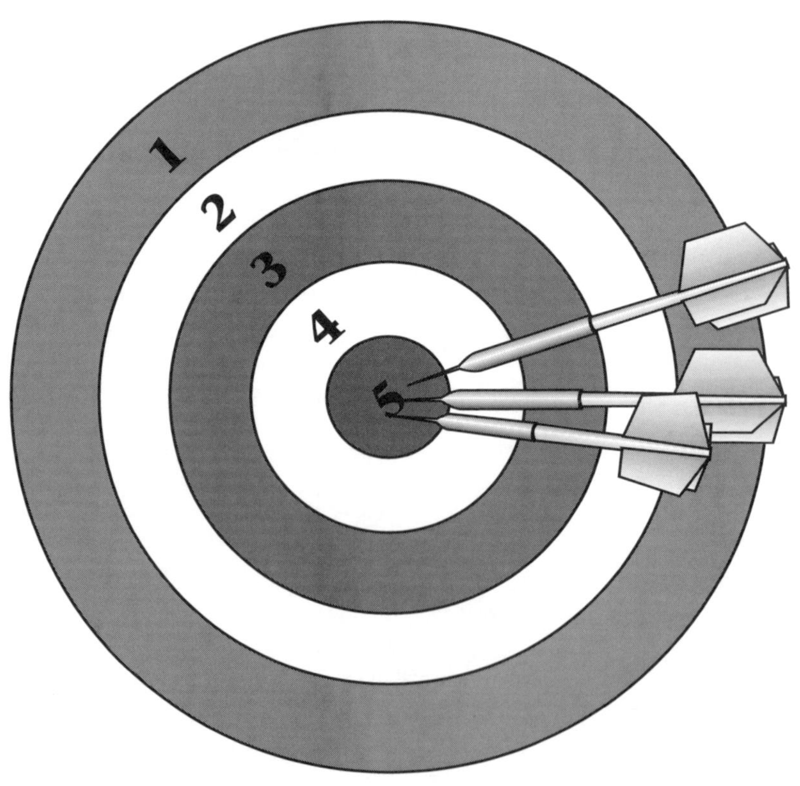

The areas of each circle: ____

____

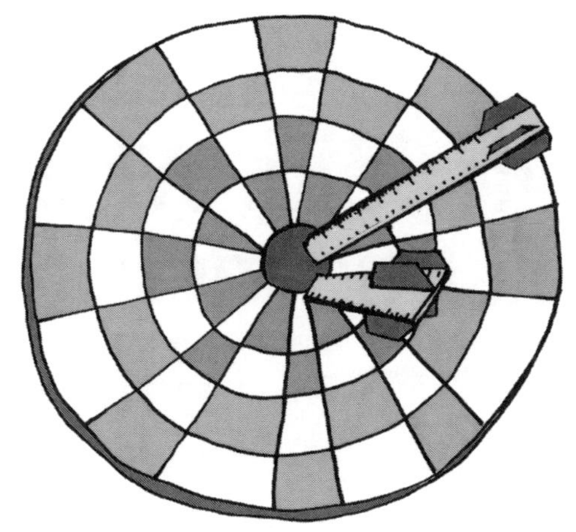

The Dartboard: What Are the Chances?

Game Sheet

Our Game—Points and Rules

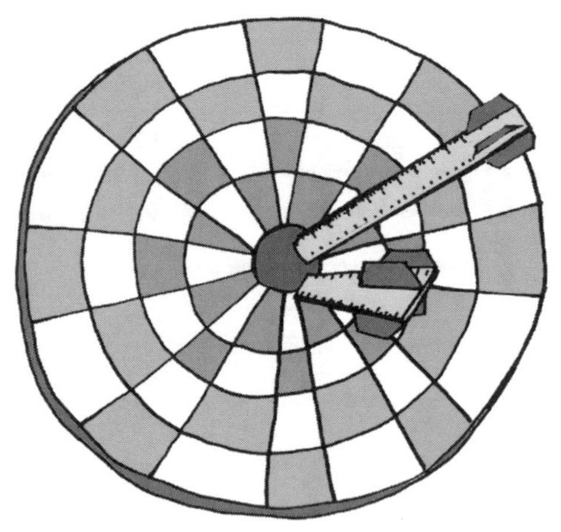

The Dartboard: What Are the Chances?

Grading Matrix

Names ____________________

Date__________ Class ____________

Criteria	4	3	2	1
Accuracy of measurement				
Quality and accuracy of computation (area of concentric circles)				
Caliber of game rules				
Overall quality of written explanation				
How well did group work together?				

Comments:

4 = Superlative, 3 = Competent, 2 = Limited, 1 = Inadequate

ACTIVITY 17

Math and Music: Frequencies

MATH TOPICS

Ratios, Measurement, Computation, Rounding, Problem Solving

TYPES OF INTELLIGENCES

Musical/Rhythmic, Logical/Mathematical, Visual/Spatial, Verbal/Linguistic, Interpersonal

CONCEPTS

Students will do the following:

1. Investigate frequencies in the C major scale.
2. Find the ratios between the frequencies of each note.
3. Problem solve a relationship between these ratios.
4. Work in a group of four to build a panpipe using these ratios to calculate the length of each pipe.

MATERIALS

- Copies of Activity Sheet 1 and 2 for each student
- Copy of Activity Sheet 3 for each group

- Copy of Music Sheet for each group
- Copy of Grading Matrix for each group
- $^7/_8$″ PVC pipe for each group
- Pipe cutter (to be used by an adult)
- File to smooth pipe (one for each group)
- String
- Calculators

WHAT TO DO

Give each student copies of Activity Sheet 1 and Activity Sheet 2 and discuss the concept of frequencies. Ask, "What appears to be the relationship between notes that are one octave apart? For example, middle A is 440 vib/sec, the A below middle A is 220 vib/sec, and the A above middle A is 880 vib/sec. What is the relationship? What do you notice about the frequencies of middle C and the C above middle C?" Students should notice that the same note one octave higher has twice the frequency. The ratios are shown in the following table. They are listed in the column relating to the numerator of the ratio.

Scale Note	C	D	E	F	G	A	B	C
Frequency (vib/sec)	264	297	330	352	396	440	495	528
Ratio of Frequencies		1.125	1.111	1.067	1.125	1.111	1.125	1.067

Once students have determined these ratios and understand the relationship between the notes, they can use this information to determine the lengths of the pipes for the panpipes they will be building. It should be noted that when a black key separates two notes, for example, C and D, it is considered a whole step, and when there is no black note between them, for example, between E and F, it is considered a half step.

Distribute copies of Activity Sheet 3 and Music Sheet to each group. Explain how the lengths of each pipe have been calculated: Middle C is measured at 8″; $D = 8 \div 1.12 \approx 7^1/_8$;

$E = 7^1/_8 \div 1.12 \approx 6^3/_8$; etc. The following table shows the complete list of measurements. Discuss with the class the divisors used to calculate the length of each pipe. Since there are no black keys between E and F and B and C, these are half steps. Instead of dividing by 1.12, students divide by 1.06. Have students discuss this numerical relationship (.06 is half of .12).

Note	Divided by	Length of Pipe (rounded to the nearest $^1/_8$″)
C	—	8″
D	1.12	$7^1/_8$″
E	1.12	$6^3/_8$″
F	1.06	6″
G	1.12	$5^3/_8$″
A	1.12	$4^3/_4$″
B	1.12	$4^1/_2$″
C	1.06	4″

VARIATION

Students can explore the frequencies of notes in other scales to discover their relationships.

TIE TO TECHNOLOGY

Commercial programs and software are available that allow musicians to compose music using the computer. Some students who are interested in and have studied music may have access to one of these programs and can be encouraged to share this technology with the class.

Electronic synthesizers have become a very popular way to create music. It has been said that musicians are limited by the fact that they have "only two hands" while the synthesizer can be used to create the music of an entire band. Synthesizers can be made to communicate with each other (via computer) with the help of sound files created by MIDI (Musical Instrument Digital Interface). Using the keyword *MIDI,* students can locate and download information from the Internet to produce music on their computers. (See appendix for specific sites and their URLs.)

ASSESSMENT

1. Student products
2. Journal question: "Do you think the diameter of the pipe would affect the sounds produced by the panpipe? Explain your answer."
3. Grading matrix

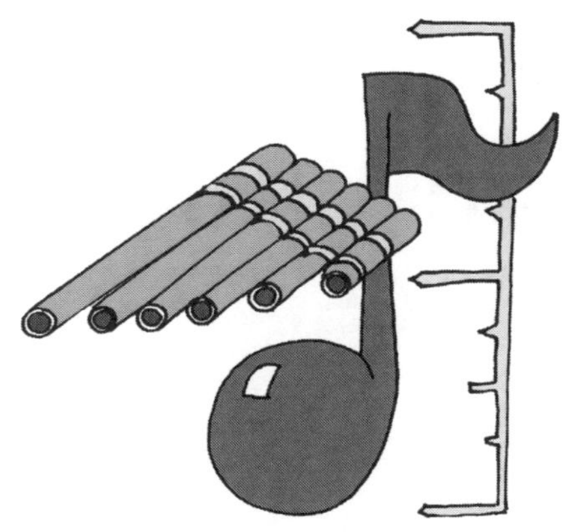

Math and Music: Frequencies

Activity Sheet 1

The sounds produced from a musical instrument are the result of vibrations. Moving a string (guitars or pianos) or moving a column of air (piccolo or flute) can cause the vibration or the vibration may be the result of moving air over a reed (oboe or French horn). The frequency or number of vibrations per second determines what pitch the note has. If a note vibrates at 440 vibrations per second (vib/sec), it produces a sound heard as a middle A. The note that is one octave below middle A has a frequency of 220 vib/sec, and the A above middle A has a frequency of 880 vib/sec. The following keyboard shows the octave from middle C to the next C above it. The table indicates the frequency of each of these notes.

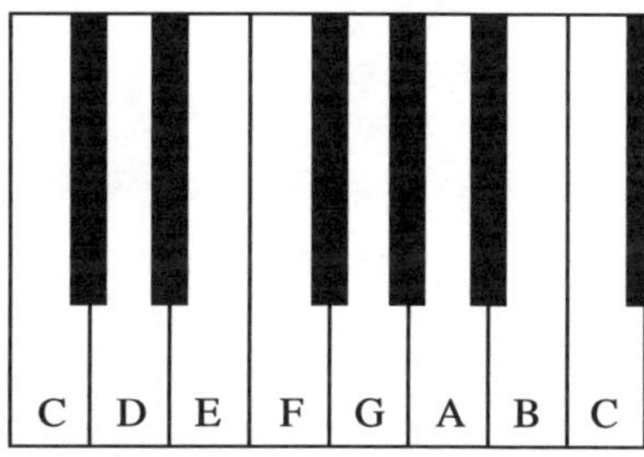

Scale Note	C	D	E	F	G	A	B	C
Frequency (vib/sec)	264	297	330	352	396	440	495	528
Ratio of Frequencies		1.125						

Find the ratio between each of the frequencies. For example:

$$\frac{D}{C} = \frac{297}{264} = 1.125$$

Find the other ratios and list them on the table. What relationship do you see?

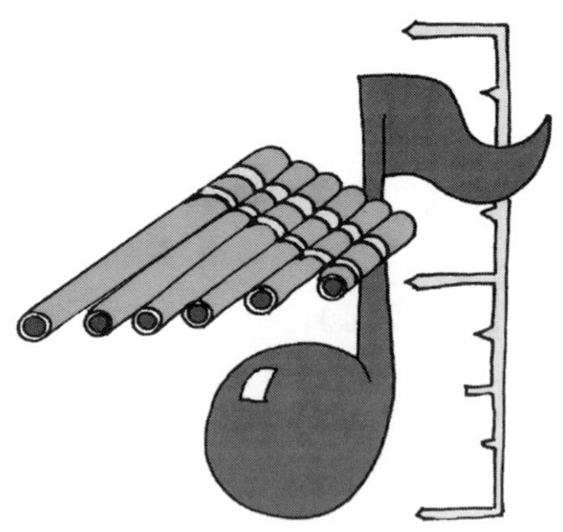

Math and Music: Frequencies

Activity Sheet 2

We can use our newly found understanding of frequencies to build a real musical instrument—a panpipe! The length of each of the pipes determines the frequency of each note. A panpipe looks like this:

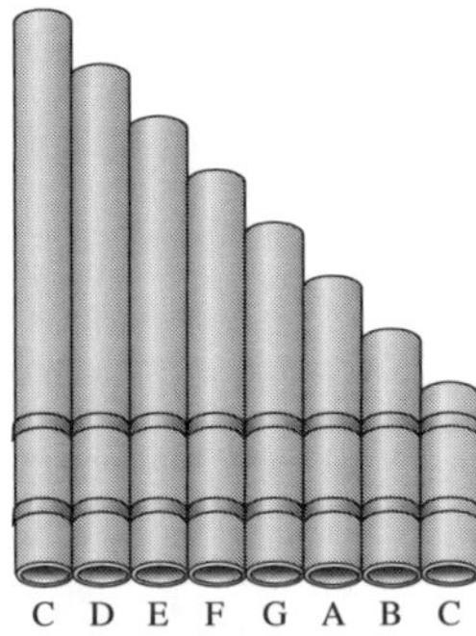

The longer the pipe, the deeper the tone. Middle C, with a frequency of 264 vib/sec, is the longest of the pipes.

You will be working with your group to build a panpipe like this one. The length of each pipe is shown on the following table. Do you see any relationship between the lengths of the pipes and the frequency ratios you calculated earlier? What is this relationship?

Note	Divided by	Length of Pipe (rounded to the nearest $^1/_8$″)
C	—	8″
D	1.12	$7^1/_8$″
E	1.12	$6^3/_8$″
F	1.06	6″
G	1.12	$5^3/_8$″
A	1.12	$4^3/_4$″
B	1.12	$4^1/_2$″
C	1.06	4″

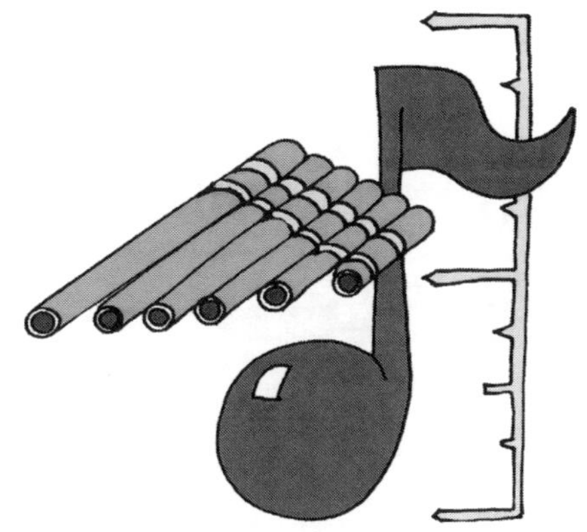

Math and Music: Frequencies

Activity Sheet 3

Directions

Work in groups of four.

1. Calculate the total length of pipe you need to construct the panpipe.
2. When you get the overall length you need, mark off the length of each individual piece and have the adult helper cut the pipe for the group.
3. Use the file to sand the pipe ends—the ends must be very smooth as you will be blowing over the edge of the pipes to produce the sounds.
4. Work together to connect the pipes in the correct places using the length of string—you must problem solve a way to keep the pipes steady and in place.
5. Use the panpipe to compose an original piece of music. Be sure to write the music correctly on the music sheet provided. It must be at least eight measures long.
6. If you have time, write lyrics to help understand a mathematics concept (model after a math medley).

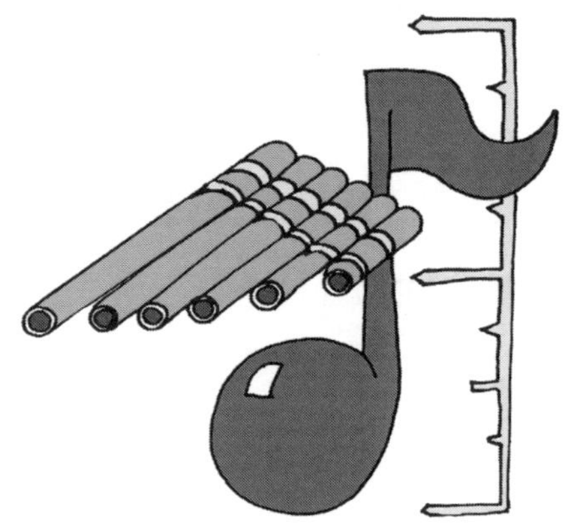

Math and Music: Frequencies

Music Sheet

Directions

Use the musical staffs below to compose your original musical composition for the panpipes. Be sure to develop a time signature and eight measures. Use the correct notes and symbols.

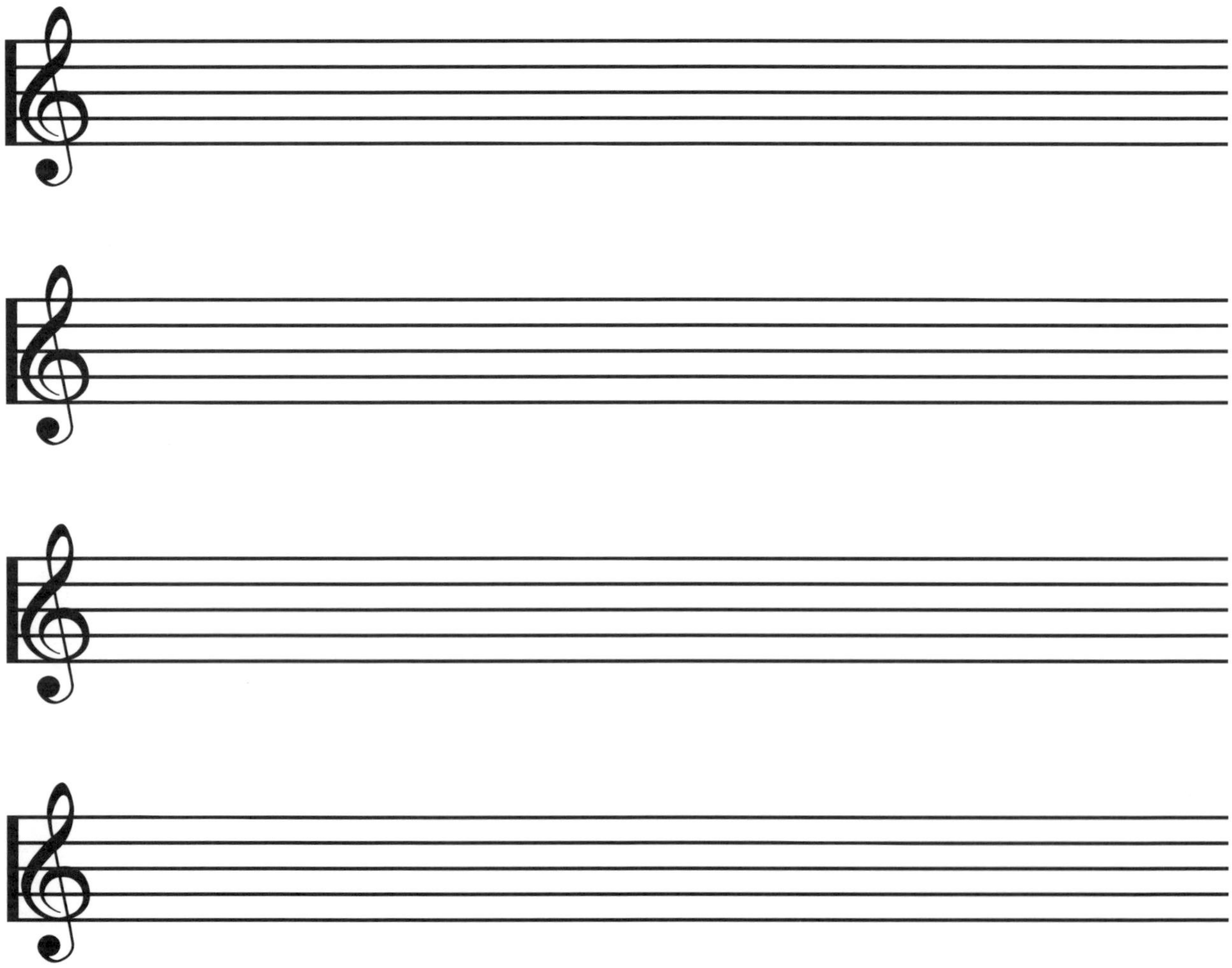

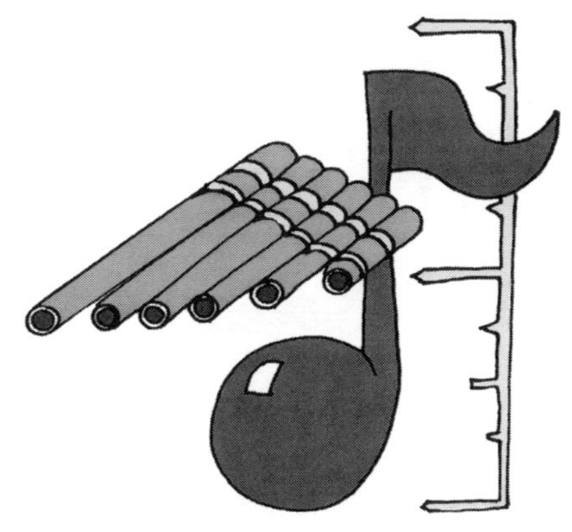

Math and Music: Frequencies

Grading Matrix

Names ______________________________

Date____________ Class ______________________

Criteria	4	3	2	1
Accuracy of computation				
Analysis of frequency ratios				
Measurement and design of panpipe				
Quality and originality of song				
How well did group work together?				
Comments:				

4 = Superlative, 3 = Competent, 2 = Limited, 1 = Inadequate

Paper-Folding Polygons

MATH TOPICS

Informal Geometry, Angle Measurement, Central Angles, Triangles, Symmetry, Problem Solving

TYPES OF INTELLIGENCES

Bodily/Kinesthetic, Logical/Mathematical, Verbal/Linguistic, Visual/Spatial

CONCEPTS

Students will do the following:

1. Follow oral and written directions to fold a paper into a pentagon.
2. Analyze shapes and angles.
3. Explore the concept of symmetry.

MATERIALS

- Copy of Directions Sheet for each student
- Copy of Grading Matrix for each student
- One $8^1/_2'' \times 11''$ sheet of paper for each student
- Scissors
- Markers, colored pencils, or crayons
- Protractors and rulers

WHAT TO DO

Explain to students that they are going to start this project with a standard sheet of $8^1/2'' \times 11''$ sheet of paper and end with an unusual five-sided snowflake. Explain that as the folding takes place you will be asking them questions about the shapes and angles that are being formed and they must pay close attention and problem solve the answers before they continue.

The more precise the folds, the more accurate the measurements will be. Step-by-step directions for folding are on Directions Sheet. Students should be asked to measure the angles that are being formed at E in steps 4, 5, and 6. Have students measure ∠E when they have completed the folding operation. The angles formed at this vertex are the central angles of the pentagon—there are ten of them and they all should be congruent. They are 36°.

At various times during the folding, right, obtuse, and acute triangles are formed. Many of the right triangles are similar with angles of 36°, 90°, and 54°.

Once students have folded the paper to the smallest shape desired (or, in other words, have made the final fold), ask them to draw a design on the exterior of the paper and cut out the design (the blackened areas have been cut out). Ask students to predict what the design will look like once they open the paper. Examples of both are shown here:

When the triangle-shaped paper is opened, the pentagonal design to its right is formed. It is an interesting activity for the students to analyze the shape before and after the paper is opened.

VARIATION

Other geometric shapes can be paper-folded. Hexagons can be used to decorate winter windows (these are true snow-flakes), and octagons are an interesting project.

TIE TO TECHNOLOGY

Dynamic geometry software can be used to help students draw and analyze various polygons.

ASSESSMENT

1. Student products
2. Observation and questioning: During the paper-folding activity, ask students about the types of triangles being formed, the size of the angles, the central angles, the interior angles, etc.
3. Journal question: "Explain the difference between the interior angles of the pentagon and its central angles. What must the sum of the central angles of any polygon be and why?"
4. Grading matrix

Paper-Folding Polygons

Directions Sheet

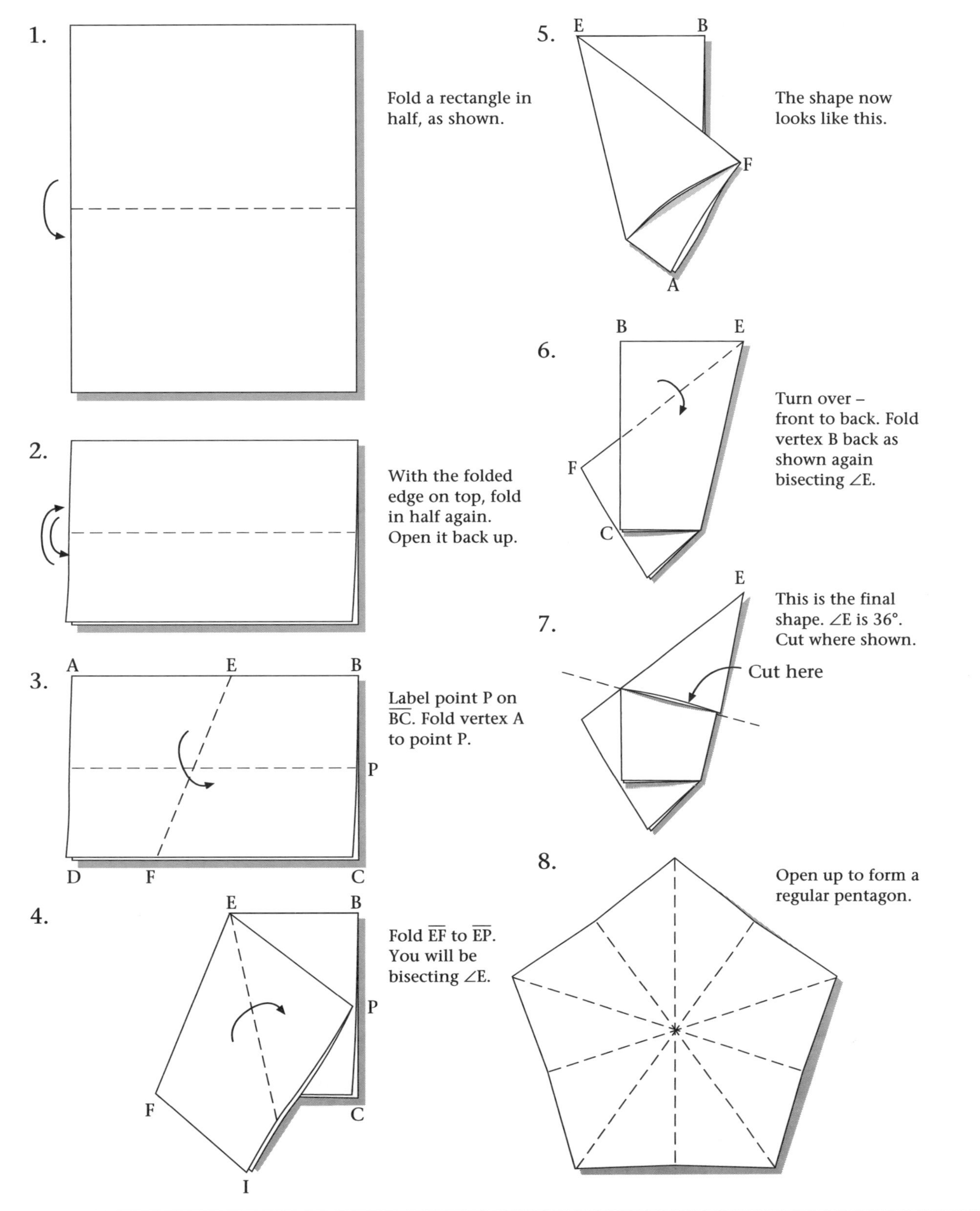

Paper-Folding Polygons

Grading Matrix

Names ____

Date____________________ Class ___________________________

Criteria	4	3	2	1
How well did student follow directions?				
Analysis of shapes and angles				
Overall accuracy of paper-folding				
Quality and originality of symmetrical design				
Comments:				

4 = Superlative, 3 = Competent, 2 = Limited, 1 = Inadequate

ACTIVITY 19

The Valley of Mars

MATH TOPICS

Measurement, Estimation, Appropriate Units of Measure, Computation, Averages, Conversions, Open-Ended Problem Solving

TYPES OF INTELLIGENCES

Logical/Mathematical, Bodily/Kinesthetic, Interpersonal, Verbal/Linguistic, Visual/Spatial

CONCEPTS

Students will do the following:

1. Problem solve a method using cookies as a unit of measurement to measure large distances.
2. Use their measurements to calculate the number of cookies needed to span a specific distance.
3. Calculate the amount of money needed to buy the required quantity of cookies.
4. Compare their estimates with those of other groups to find a class average (for both number of cookies and cost).
5. Use conversions to change cookies/cm to cookies/km.

MATERIALS

- Copy of Activity Sheet for each group
- Copy of Grading Matrix for each group
- Overhead transparency of Class Data Sheet
- Five cookies for each group of four students (Oreo cookies work well because they are quite uniform in diameter)
- Metric rulers
- Calculators

WHAT TO DO

Discuss with students the concept of unit of measure. Ask them to name some standard units used to name measurements. They might suggest terms such as *inches, feet, meters,* and *kilometers.* Ask students if they have ever used any less traditional or nonstandard units; for example, the size of their foot. Very often, young children will measure a length using paper clips or estimate the weight of an object as being "heavier than this apple." Most likely, students will have experience using both standard and nonstandard units of measurement.

Distribute one Activity Sheet to each group and ask students to predict how many cookies long the Mariner Valley on Mars might be. Ask them to write their estimates so they can be referred to at the end of the lesson.

It is a good idea for students to measure more than one cookie because not every cookie is exactly the same size. Once the students have their measurements, they will need to convert in the following manner:

cookies/cm → cookies/m → cookies/km → cookies/4,537 km

To calculate the cost of the cookies, students will need to know the cost of one package of cookies. Surprisingly, it will cost hundreds of millions of dollars (depending on the cost of the bag or box you supply).

When all groups have completed their calculations, ask them to enter their estimates for length and cost on the overhead transparency of the Class Data Sheet. Once the average or mean has been computed, students can calculate other data, as described in the Variation section.

VARIATION

Class means and individual estimates can be used to calculate range, median, mode(s), variation from the mean, and other statistical information.

ASSESSMENT

1. Traditional tests and/or quizzes on conversions.
2. Observation and questioning: "Explain how you set up the problem to help you convert units of measurement."
3. Journal question: "Why was it necessary to measure more than one cookie when conducting this experiment?"
4. Grading matrix

ON THE INTERNET

Two astronomical sites with breathtaking pictures that students can use in a PowerPoint presentation are Welcome to the Planets and NASA's Planetary Photojournal (see appendix for URLs). Both sites have awe-inspiring pictures of all of the planets and show their relative size in its opening graphic. NASA's Planetary Photojournal provides a viewing option—users can see pictures as if they were looking through the Hubble telescope or see pictures taken on space voyages. Amazing! At both sites, students can access all of the planets for research purposes.

The Valley of Mars

Activity Sheet

The Grand Canyon, in Colorado and Arizona, is about 349 km long, but Mars has the longest canyon in the solar system. It is called the Mariner Valley and is 13 times longer than the Grand Canyon. It is 4,537 km long, and if it were placed in the United States, it would stretch from the east to the west coast.

What if we placed round cookies end-to-end and used these cookies as our unit of length? How many cookies would it take to measure the length of the Mariner Valley on Mars? And how much would all of these cookies cost?

Directions

Work with your group to solve this problem.

1. Use 5 cookies to help you solve this problem.
2. Remember, you are measuring in centimeters, but you must convert these measurements to kilometers to find the answer to this problem.
3. Calculate the cost of the total number of cookies needed.

Show your work here.

The Valley of Mars

Class Data Sheet

Group	Number of Cookies	Cost of Cookies
Average		

Were the class estimates very accurate? If not, why do you think it was so difficult to estimate the number of cookies and their cost? Discuss these questions with your group and assign a spokesperson to explain your thinking.

The Valley of Mars

Grading Matrix

Names ____

Date____________________ Class ______________________________

Criteria	4	3	2	1
Accuracy of computation (number of cookies)				
Accuracy of computation (cost of cookies)				
Quality of problem-solving strategies				
How well did group work together?				
Comments:				

4 = Superlative, 3 = Competent, 2 = Limited, 1 = Inadequate

ACTIVITY 20

Limitless Lung Power

MATH TOPICS

Three-Dimensional Geometry, Measurement, Circumference of a Circle, Volume of a Sphere, Mean, Deviation from the Mean, Collection, Organization, Analysis of Data, Estimation

TYPES OF INTELLIGENCES

Interpersonal, Intrapersonal, Logical/Mathematical, Verbal/Linguistic, Bodily/Kinesthetic

CONCEPTS

Students will do the following:

1. Measure and use the circumference of a balloon to calculate its diameter and radius.
2. Use the average circumference to find the lung power of the group.
3. Make use of class's predictions to analyze their data using deviation from the mean.
4. Work collaboratively to problem solve.

MATERIALS

- Copy of Activity Sheet for each group
- Copy of Grading Matrix for each group
- Overhead transparencies of Class Predictions and Class Data Chart
- Basketball
- One balloon for each student
- Tape measure for each group
- Calculator for each group

WHAT TO DO

Ask students, "How much air do you think it takes to inflate this basketball?" Measure the circumference with a tape measure to the nearest centimeter. It should be between 74 cm and 78 cm (depending on how inflated the basketball is).

For demonstration purposes, assume the circumference is 76 cm. Demonstrate the procedure used to find the volume of a sphere. Ask students what formula is used to find the circumference of a circle. (The formula is $c = \pi d$.) Using algebra transformations, $c/\pi = d$. Using 76 cm for the circumference and 3.14 for π ($76/3.14 = d$), we find the diameter (d) is approximately 24 cm. To calculate the radius, divide the diameter (24) by 2; the radius of the basketball is approximately 12 cm. Students now have all of the information they need to find the volume of a sphere (e.g., the basketball).

The formula used to find the volume of a sphere is $V = 4/3\pi r^3$. Substituting into the formula, $V = 4/3 \times 3.14 \times 12^3$. Using order of operations, multiple exponents first: $12^3 = 1{,}728$; $4/3 \times 3.14 \times 1{,}728 = 7{,}234.56\ cm^3$ or approximately $7{,}235\ cm^3$.

Ask students to estimate how much "hot air" their balloon will hold (based upon what they know about the volume of the basketball). Write their predictions on the overhead transparency Class Predictions. Explain that these predictions will be evaluated after the experiment is conducted.

Have students form into groups of four and give each student a balloon. Have them stretch it out and blow into it a few times to loosen it up. Give each group a calculator, tape measure, and Activity Sheet. Explain that the scientific process suggests that more than one trial is necessary to make the data more accurate. Each group should find its group average. Ask, "What will this number represent?" Groups will use this average to find the mean (or average) volume of their balloons—the group's "hot air."

Show the overhead transparency Class Data Chart and ask each group to write its average lung capacity to the nearest centimeter. When all groups have completed the task, use the class predictions to analyze the deviation of the data from the class mean.

VARIATION

Students can find the volume of irregularly shaped three-dimensional objects by using the displacement of water. Place water in a rectangular prism and find its volume. Place the object in the water and determine the new volume—the water has risen because of displacement. The increase in water volume is the volume of the irregularly shaped object.

TIE TO TECHNOLOGY

The volume of assorted polyhedra can be calculated by using a spreadsheet. The following spreadsheets show examples of rectangular prisms and pyramids by both numbers and formulas.

	A	B	C	D	E
1	**Comparing the Volumes of Rectangular Prisms and Pyramids**				
2					
3	**Length**	**Width**	**Height**	**Volume of Prism**	**Volume of Pyramid**
4	8	6	5	240	80
5	10	12	8	960	320
6	7	4	9	252	84
7	3	6	2	36	12
8					

	A	B	C	D	E
1	**Comparing the Volumes of Rectangular Prisms and Pyramids**				
2					
3	**Length**	**Width**	**Height**	**Volume of Prism**	**Volume of Pyramid**
4	8	6	5	=A4*B4*C4	=(1/3)*D4
5	10	12	8	=A5*B5*C5	=(1/3)*D5
6	7	4	9	=A6*B6*C6	=(1/3)*D6
7	3	6	2	=A7*B7*C7	=(1/3)*D7
8					

ASSESSMENT

1. Student products
2. Observation and questioning: "What is the range of your group's air capacity? Do group members appear to have similar lung capacities or does one person have a much greater or much lower lung capacity? What might account for these outliers?"
3. Journal question: "You have a sport's ball with a circumference of 24 cm. Find its volume. Explain your answer."
4. Grading matrix

Limitless Lung Power

Activity Sheet

Has anyone ever told you that you're full of hot air? Well you are, you know! We all have a great deal of hot air in our lungs or we'd be in a lot of trouble. How much air do you think you have in your lungs? Remember that we're dealing with volume and so we are measuring in cm^3. Use what you know about the volume of the basketball to help you make your predictions. Make your predictions now. All predictions will be recorded on the Class Predictions transparency.

Directions

Work in groups of four to measure your limitless lung power.

1. Each person is to take turns doing the following tasks, so that each person performs all of the tasks:
   - Blowing up the balloon.
   - Measuring its circumference (to the nearest mm).
   - Recording the measurements.
   - Calculating the average circumference.
2. Record your results in the data collection table shown below.
3. Each person is to make 3 trial runs and find the average of all 3 trials.
4. Use the easy-to-follow steps on the worksheet to find your lung power and the average for your group.
5. Record the lung power of the group (lung capacity) on the overhead transparency of the Class Data Chart.

Data Collection Table

Person	Circumference Trial 1	Circumference Trial 2	Circumference Trial 3	Average Circumference
Average for Group				

6. Now, use the group average to calculate the mean lung capacity (lung power) of the group. Since the balloon is a sphere, use the formula $V = {}^{4}/_{3}\,\pi r^3$ and follow these simple steps:

a. To find the radius for this formula, use the average circumference to calculate the diameter. Since $C = \pi d$, then $d = C \div \pi$. (**Note:** Use 3.14 for π.)

b. To calculate the radius, divide the diameter in half: $4 = d \div 2$.

c. Substitute the radius into the formula for finding the volume of a sphere to find the average (mean) lung power of your group.

d. Record your average on the Class Data Chart.

Our Work:

Limitless Lung Power

Class Predictions

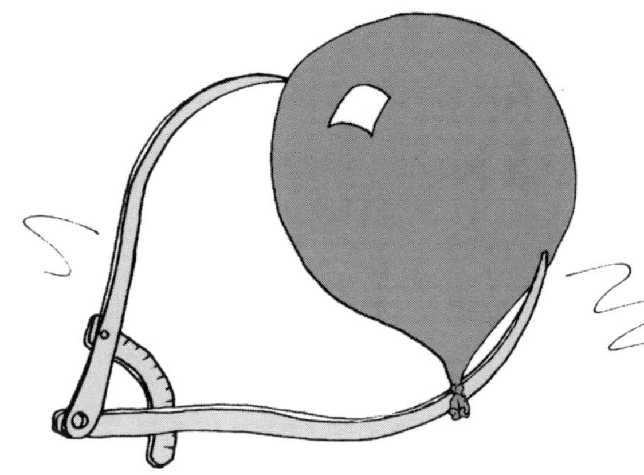

Limitless Lung Power

Class Data Chart

Group	Average Lung Capacity
Class Average	

How accurate were your predictions? Go back to the prediction sheet and check them for the following:

1. How many predictions were above the mean?
2. How many predictions were below the mean?
3. Which predictions were the furthest from the mean?

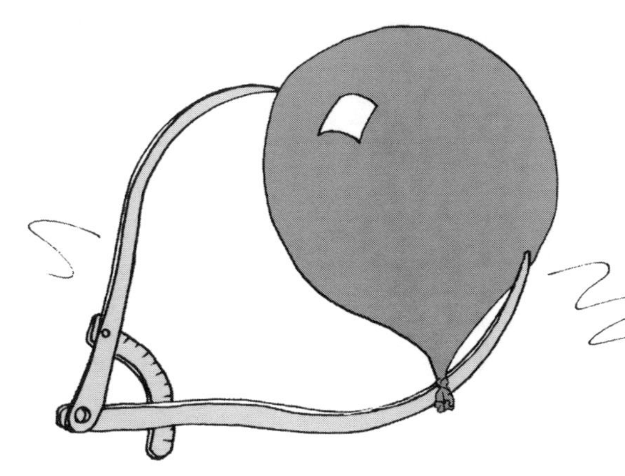

Limitless Lung Power

Grading Matrix

Names ______________________________

Date______________ Class ______________________

Criteria	4	3	2	1
Accuracy of measurement				
Accuracy of computation				
How well did group work together?				
Comments:				

4 = Superlative, 3 = Competent, 2 = Limited, 1 = Inadequate

ACTIVITY 21

Tessellations: Transformations a la Escher

MATH TOPICS

Transformational Geometry, Area

TYPES OF INTELLIGENCES

Visual/Spatial, Logical/Mathematical, Bodily/Kinesthetic, Verbal/Linguistic

CONCEPTS

Students will do the following:

1. Use translation to transform a 2″ × 2″ square.
2. Use their revised square (pattern piece) to tile the plane.
3. Color their designs to create mathematical artwork.

MATERIALS

- Copy of Sheet of Squares for each student
- Overhead transparency of Sheet of Squares
- Copy of Activity Sheet for each student
- Copy of Grading Matrix for each student
- Paper for each student

- 3″ × 5″ inch index cards (one for each student)
- Scissors and glue sticks
- Markers or colored pencils
- Scotch tape (teacher will need a roll of clear tape to use on the overhead; students can use any kind of tape)

WHAT TO DO

Some Background Information: A tessellation is a tiling of the plane. There are regular tessellations—tiling the plane with only one polygon. Regular tessellations can be created using triangles, quadrilaterals, and hexagons because the sum of the central angles at each vertex is 360°. Another type of tessellation, semiregular, is accomplished by using two or more regular polygons. There are eight different semiregular tessellations.

Islamic artists are the masters of tessellation. Islam forbids the making of images and so all of their artwork is produced using geometric shapes only. A Dutch artist, M. C. Escher, studied and sketched the drawings on the walls and ceilings of the Moorish palace, the Alhambra, and spent many years learning to create his now-famous nonpolygonal tessellations of birds, fish, reptiles, and people. In this lesson, students will use squares to create their own works of art—a la M. C. Escher.

Discuss this background information with students and explain that they are going to create some of their own "works of art."

Give each student a copy of Sheet of Squares and have them cut out a 2″ × 2″ inch square. Ask them to discuss the area of this shape; ask what the perimeter is. Show an overhead transparency of Sheet of Squares and use these squares to demonstrate the technique of designing a tessellation. The students are going to create a nonpolygonal tessellation puzzle piece using translations (or slides). This is the simplest of the transformations and the best one to introduce the topic.

Use the example provided on the Activity Sheet to show how the square is transformed into a "friendly fish." Explain that the lines on the square can be used to ensure that transformations are placed correctly. Work along with the students to change one side of the square, cut it out, and physically move it to the opposite side. Be sure to align it properly. Use the clear tape to affix it to that side. Now ask students to design their own change, cut it out, and move it to the opposite side. Do the same with the third side—translating it to the opposite side.

When students have completed the transformation, they must glue their "puzzle piece" to the index card and carefully cut it out again. Explain that great care should be used when cutting to ensure that the piece will tessellate, or tile, the plane.

VARIATION

Students can use triangles, quadrilaterals, and/or hexagons to make their puzzle pieces. They can also try other types of transformations—rotations and reflections.

TIE TO TECHNOLOGY

A most creative software program for creating computer-generated tessellations is TesselMania. Produced by MECC, it allows students to use a triangle, quadrilateral, or hexagon as the base figure and to perform a variety of transformations on the figure. They then can color in the tessellation to produce an original work of art. This software is readily available through many educational catalogues.

ASSESSMENT

1. Tessellations
2. Journal question: "Explain why the tiling produced by your nonpolygonal tessellation is called a translation or a 'slide.' What do you think the tiling would look like if we rotated our 'squiggles'?"
3. Grading matrix

Tessellations: Transformations a la Escher

Sheet of Squares

Tessellations: Transformations a la Escher

Activity Sheet

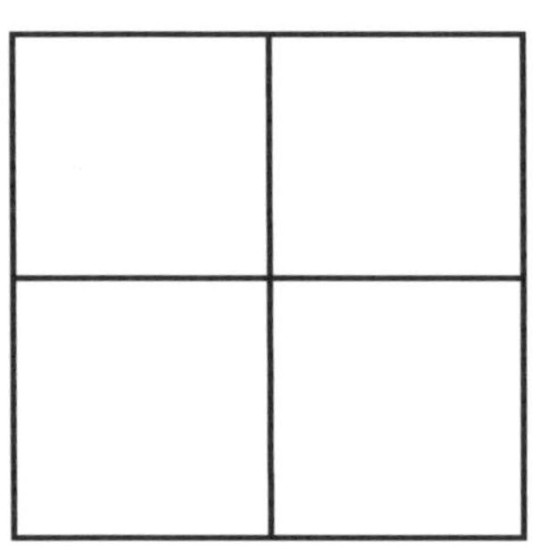

1. Start by cutting a 2″ × 2″ square.

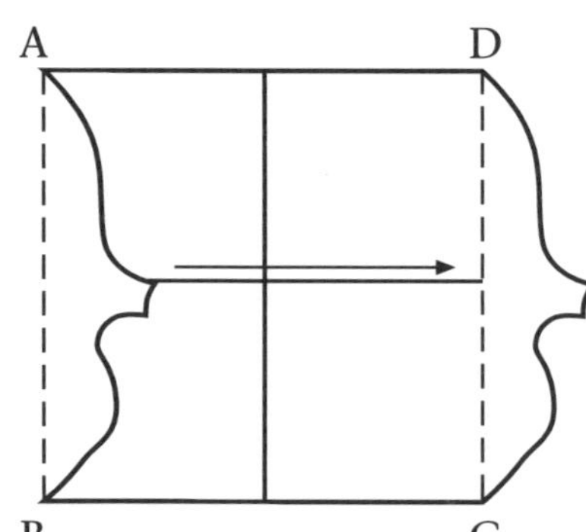

2. Cut a "squiggle" from $\overline{AB}$ and "slide" it to the opposite side ($\overline{CD}$).

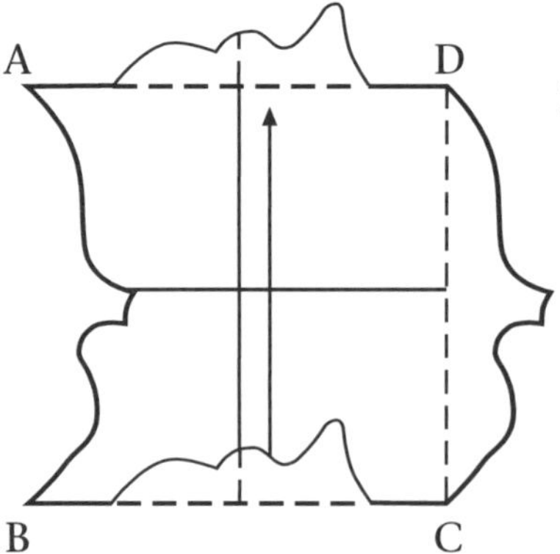

3. Cut a "squiggle" from $\overline{BC}$ and "slide" it to the opposite side ($\overline{AD}$).

4. You have completed your nonpolygonal tessellation. Use your imagination. What does it look like?

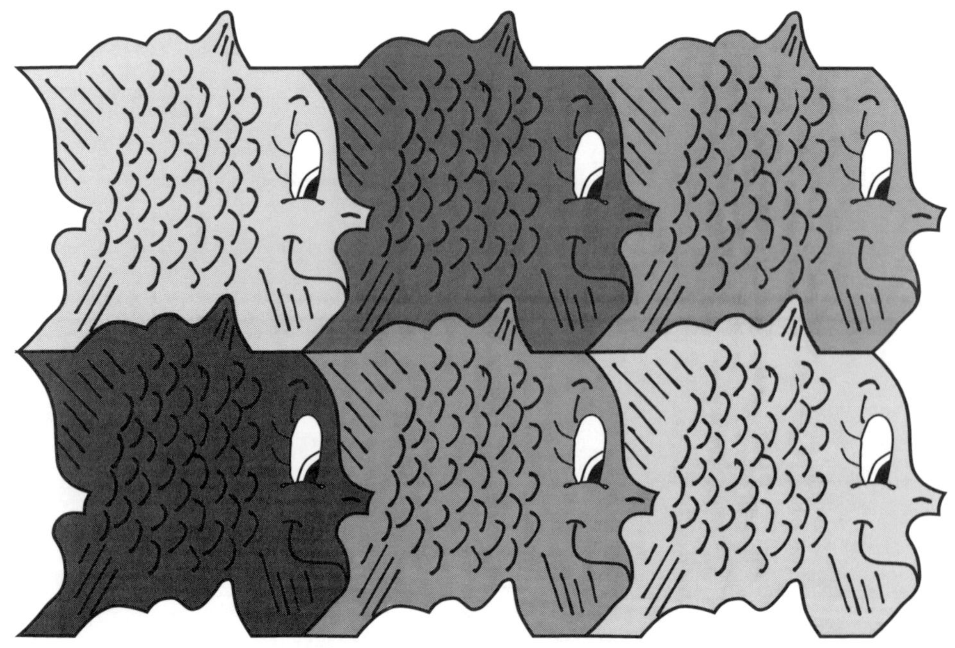

5. Use color and design to "tile the plane." A complete tessellation a la M. C. Escher.

Tessellations: Transformations a la Escher

Grading Matrix

Names ______________________________

Date________________ Class ______________________________

Criteria	**4**	**3**	**2**	**1**
How well did student follow oral directions?				
Accuracy of measurements				
Overall quality of tessellation				
Comments:				

4 = Superlative, 3 = Competent, 2 = Limited, 1 = Inadequate

CHAPTER 4

Data Analysis, Statistics, and Probability

CHAPTER 4

Data Analysis, Statistics, and Probability

Statistical thinking will one day be as necessary for efficient citizenship as the ability to read and write.—H. G. Wells (1866–1946)

For students to understand the importance of statistics and data collection in our information age, activities and projects should be designed that model how statistics are used in the real world. All too often students are asked to analyze textbook data and solve meaningless problems. To develop worthwhile experiences for students, teachers must investigate the reasons data is collected. In the real world, people collect data to

- Make predictions.
- Find mathematical constants.
- Sample populations that are too large to count.
- Help explain patterns or trends.
- Answer questions regarding preferences.
- Help explain natural phenomena.

The first activity, Predicting Colors in a Bag of M&Ms, involves students in a very tasty experiment that demonstrates the usefulness of data collection and analysis in making predictions. How confident would anyone be in guessing how many of each color M&Ms there are in a mystery bag? In this activity, students collect data that allows them to answer this question with an informed estimate rather than an uninformed guess.

Phone Home is an amusing data collection activity that affords students the opportunity to experience the speed of light, a natural phenomenon, and explain its speed in concrete terms. Miranda, the Martian is homesick and sends a message home. How long will it take, at the speed of light, for the message to reach her loved ones? By conducting this experiment, students can approximate the amount of time it would take this message to travel through space from Earth to Mars.

Have you ever wondered, when filling out a form that contains little boxes or spaces to write individual letters and numbers, how the designers of the form determined how many spaces (or boxes) to allow for each item? For example, how did they determine how many spaces to allow for the letters in a person's first name? Why do most people not run out of spaces? The activity How Long Is Your First Name? uses statistics and deviation from the mean to help students answer this question. It is a motivating approach to explain patterns, or trends.

We, as humans, look somewhat alike because we are proportional. It is possible to develop mathematical constants when comparing the relationship, or ratio, between various body measurements. Artists have been using these ratios for centuries to produce what is considered art. Shoe Length versus Height gets students actively involved in collecting just such a mathematical constant—the relationship between the size of one's shoe and one's height. While there are some exceptions, generally speaking, the larger a person's shoe size, the taller the person is. This relationship, or correlation, is addressed when the data is graphed on a scatterplot and the coordinates on the graph are analyzed.

The Millions of Marshmallows activity will intrigue students because it answers the question, "How do we count objects when there are too many to actually count?" For example, how do we know how many whales there are in the North Atlantic? Do they all stand still while scientists run (or swim) around counting them? Students learn that the statistical technique of sampling makes it possible to approximate populations that can't actually be counted. By "tagging" some marshmallows (substituting colored ones for white ones) and taking samples, students predict the number of marshmallows in their group's bag. Another tasty way to teach statistics!

And the final activity gets students immersed in The Case of the Disappearing Telephone Numbers. Just how many telephone numbers are available, and why are we suddenly running out of them? In the first part of this activity, students explore the combinations of available numbers. In the second part, they attempt to answer the question of why the available numbers are vanishing so quickly.

After students collect their data, they can enter it into spreadsheet programs and use a computer to analyze the data and make appropriate graphs. Students can experiment with a variety of graphs to find the one(s) that best portray the data.

Each of the activities and projects in this chapter has been designed to relate the study of statistics and data collection to the role they play in our modern world. By making connections between school mathematics and real-world math, students are more motivated because they understand the relevance of what they are learning.

ACTIVITY 22

Predicting Colors in a Bag of M&Ms

MATH TOPICS

Collection, Organization, and Analysis of Data, Box-and-Whisker Plots, Percent of Difference, Absolute Value, Problem Solving, Critical Thinking

TYPES OF INTELLIGENCES

Logical/Mathematical, Verbal/Linguistic, Interpersonal, Bodily/Kinesthetic

CONCEPTS

Students will do the following:

1. Use data collection to predict the number of each color of M&Ms found in a mystery bag.
2. Use class data to design a box-and-whisker plot for each color of M&M.
3. Analyze their graphs to predict the number (±1 or 2) of each color of M&Ms in the mystery bag.
4. Calculate the difference (absolute value) between the actual count and their prediction.
5. Calculate the percent of difference.
6. Use their data to analyze the manufacturing techniques of the company that makes M&Ms (Mars, Incorporated).
7. Formulate other predictions that could be made using the same data collection and graphing techniques.

MATERIALS

- Copy of Data Collection Sheet for each student
- Copy of Graphing the Data for each student
- Copy of Analyzing Data for each student
- Copy of Grading Matrix for each student
- Overhead transparencies of Data Collection Sheet
- One snack-sized bag of M&Ms for each student
- A mystery snack-sized bag of M&Ms
- Calculators

WHAT TO DO

Hold up the mystery bag of M&Ms and ask students the question posed on the first data collection sheet: "How confident would you be in predicting the number of each color of M&Ms in this particular bag? I am not interested in the total number of candies in the bag but the number of each color." As students guess numbers, ask them, "Just how confident are you in your guesses?" Explain that in the "real world," data is collected and analyzed to help answer questions such as these. By having additional information, people are able to make better estimates, or more educated guesses, and increase their levels of confidence.

Ask each student to count the number of each color of M&Ms in his or her bag and record that information on the Class Data chart. Once the data is collected, the students construct box-and-whisker plots (on the Graphing the Data sheets) by following these steps:

1. Order the data for each color from least to greatest.
2. Find the median of the data; indicate this on the appropriate spot on the number line.
3. Find the median of the lower half of the data (this is called the lower quartile); indicate this on the appropriate spot on the number line.

4. Find the median of the upper half of the data (this is called the upper quartile); indicate this on the appropriate spot on the number line.

5. Use these three median scores to form the "box."

6. Placing a dot on the numbers representing the smallest and largest data and connecting them with a line forms the whisker.

After all students have entered their data, use the class data to demonstrate this procedure using the graph supplied for the red M&Ms.

Example:

The data for a class of 20 organized from least to greatest is as follows:

0, 1, 1, 2, 2 | 2, 3, 3, 4, 4 | 5, 5, 7, 7, 7 | 8, 8, 8, 9, 9

The median (or middle score) is between 4 and 5, so the median is 4.5.

The lower quartile is 2, and the upper quartile is 7.5. These are indicated on the number line in the following way:

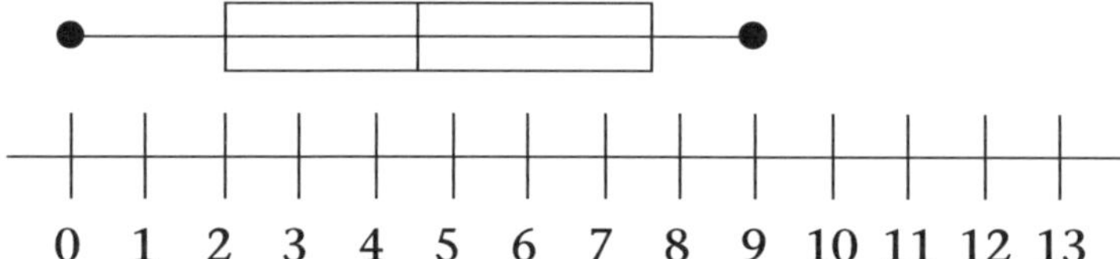

The range of the data is between 0 and 9; the median is 4.5. Fifty percent of the data falls between 2 and 7.5 (in the box).

While students still have a range of about 5 to make their predictions, the data collection and graph help them narrow down their options. For example, the prediction can be 4 ± 2 (meaning that any quantity of the specific color between 2 and 6 would be considered correct). Looking at the data, this would be a reasonable estimate.

Ask students to predict (±1 or 2) how many M&Ms of this color are in the mystery bag. For the given example, students might predict 4 (±2). Any number of M&Ms in the bag between 2 and 6 would be considered correct.

VARIATION

Students can be asked to design another data collection activity for the purposes of making a prediction about the M&Ms or another topic. Box-and-whisker plots can be designed to help analyze this data.

TIE TO TECHNOLOGY

Most graphing calculators will produce box-and-whisker plots once the data is entered. After students understand the concepts related to making these unusual graphs, they can use calculators to see how the use of technology can simplify the process.

ASSESSMENT

1. Traditional quiz is appropriate for this activity.
2. Journal question: "Describe how a box-and-whisker plot is designed. Why is it helpful in making predictions?"
3. Grading matrix

ON THE INTERNET

The manufacturer of M&Ms, Mars, Incorporated, has its own website for M&Ms (see appendix). This site provides information about every aspect of these little candies one can imagine—including what percentage of each color are manufactured.

Predicting Colors in a Bag of M&Ms

Data Collection Sheet

How confident would you be if you were asked to predict how many of each color M&Ms there were in a snack-sized bag of M&Ms? How many blue or red or orange M&Ms do you think you would find? Do you think the same number of each color would be found in each bag?

By collecting some data we can narrow our choices and more accurately predict or estimate what we might find in a "mystery" bag of M&Ms.

Directions

Count how many of each color you have in your snack-sized bag of M&Ms. Be sure to write the numbers down as you will need to report them so they can be added to the class data.

Class Data

Color	**Student Data of Number of Each Color**																			
Red																				
Orange																				
Green																				
Blue																				
Brown																				
Yellow																				

Predicting Colors in a Bag of M&Ms

Graphing the Data

Directions

Let's use the data we have collected to make one box-and-whisker plot. This type of graph will help us make more accurate predictions. First, we must take the data and order it from least to greatest and, next, use this ordered data to form a box-and-whisker plot.

Red

1 2 3 4 5 6 7 8 9 10 11 12

Orange

1 2 3 4 5 6 7 8 9 10 11 12

Green

1 2 3 4 5 6 7 8 9 10 11 12

Blue

1 2 3 4 5 6 7 8 9 10 11 12

Brown

1 2 3 4 5 6 7 8 9 10 11 12

Yellow

1 2 3 4 5 6 7 8 9 10 11 12

Our Predictions

Red? _____ Orange? _____ Green? _____ Blue? _____ Brown? _____ Yellow? _____

Predicting Colors in a Bag of M&Ms

Analyzing Data

Let's analyze the results of this experiment. Complete this table to help with the analysis.

Color	Prediction	Actual Count	Correct/Incorrect (to ±2)
Red			
Orange			
Green			
Blue			
Brown			
Yellow			
		% Correct	

1. Do you think the data collection and box-and-whisker plots helped make more accurate predictions? Why or why not? Explain your answer.

____

____

2. Mars, Incorporated says it manufactures certain percentages of each of the colors. Do you think that the same percentage of each color is manufactured? Why or why not? Explain your answer.

____

____

3. Can you think of other predictions that you could make by collecting data in this manner and then using box-and-whisker plots to find medians and ranges in the data? Describe one of these experiments.

____

____

Predicting Colors in a Bag of M&Ms

Grading Matrix

Names ______________________________

Date__________________ Class ______________________________

Criteria	**4**	**3**	**2**	**1**
Quality of box-and-whisker plots				
Accuracy of computation on analysis sheet				
Quality of answers to questions presented				
How well did group work together?				
Comments:				

4 = Superlative, 3 = Competent, 2 = Limited, 1 = Inadequate

ACTIVITY 23

Phone Home

MATH TOPICS

Data Collection and Analysis, Measurement, Conversions, Computation, Problem Solving

TYPES OF INTELLIGENCES

Logical/Mathematical, Verbal/Linguistic, Interpersonal, Bodily/Kinesthetic

CONCEPTS

Students will do the following:

1. Use a stopwatch to time the reading of a poem.
2. Find their group average of reading time to the nearest one tenth of a second.
3. Use their group average to compute the time a reading of this poem would take to travel 33.6 million miles.

MATERIALS

- Copy of Activity Sheet for each group
- Copy of The Message for each group
- Copy of Grading Matrix for each group

- Stopwatches (one for each group)
- Clipboards
- Calculators

WHAT TO DO

Discuss with the class the ideas behind the lesson. Explain that when people talk on the telephone there appears to be virtually no delay between the time a person speaks and when the person on the other end hears the speaker. But there is a delay. Because sound travels over the phone at the speed of light—186,000 miles per second—people speaking on the phone are not aware of the lag in time.

But what if the phone call is being made to a much more distant place? How long would it take then? When interviews (shown on TV) are bounced off satellites, there is a slight delay that results because of the vast distance over which sound is being transmitted.

In this activity, Miranda is an alien from Mars and she is sending a message—a poem—home. Students, working in groups of four, take turns reading the poem and recording how long it takes to read (to the nearest tenth of a second). The group's average is used to calculate how long it would take for the message to reach Mars.

VARIATION

Sound travels at a speed of 5 miles per second. Have students calculate how long it would take for the message to reach Mars at this speed.

TIE TO TECHNOLOGY

Using the graphing calculator, the individual data of the entire class can be entered into tables and a scatterplot or linear graph can be created.

Students also can use programs such as PowerPoint to prepare presentations for interdisciplinary research projects on the planets.

ASSESSMENT

1. Observation and questioning: "Could you explain one of the problems you had calculating the time the message took to get to Mars?" or "Why was it necessary to time the message to the nearest tenth of a second?"

2. Journal question: "How long would it take for your group to send the message to Mercury, which is an average distance of about 51,000,000 miles from Earth?"

3. Grading matrix

ON THE INTERNET

There are many interesting websites related to our solar system and outer space. URLs for a few of these websites can be found in the appendix.

Phone Home

Activity Sheet

We have a visitor from space! Miranda, from Mars, is really enjoying her visit but is getting a little homesick and so she is making an Inter-universal call home.

- Mars is 33,600,000 miles from Earth.
- Sounds over the telephone travel at the speed of light, which is 186,000 miles per second.

How long will it take for Miranda's message to reach Mars?

Directions

Work with your group.

1. Take turns reading Miranda's message poem.
2. Time each person using a stopwatch and record his or her time on the table.
3. Find the average time it took your group to read the message.
4. Use that average to calculate how long it took to "phone home."

Name	Time to Read Message to the nearest 1/10 second
Average Time	

We calculate the average time it will take for our message to reach home will be ______.

These are our calculations:

Phone Home

The Message

This is the message you will read and time to calculate the time it will take for Miranda to phone home!

Having a wonderful time,
Wish you were here,
Earth is really different,
And indeed quite near!

I'm sending this message,
On Tuesday, the tenth,
Please check the computer,
How much time has been spent?

We'll all work together,
To figure this out,
How long to phone home,
That's what this is about.

Use a stopwatch—keep time,
How long will it take,
For my family in space,
This rhyme to partake?

Phone Home

Grading Matrix

Names ______________________________

Date______________ Class ______________________

Criteria	**4**	**3**	**2**	**1**
Quality of data collection				
Accuracy of conversions				
Originality and caliber of problem solving				
How well did group work together?				

Comments:

4 = Superlative, 3 = Competent, 2 = Limited, 1 = Inadequate

ACTIVITY 24

How Long Is Your First Name?

MATH TOPICS

Data Collection and Analysis, Statistics, Absolute Value, Deviation from the Mean, Problem Solving

TYPES OF INTELLIGENCES

Logical/Mathematical, Verbal/Linguistic, Bodily/Kinesthetic, Interpersonal, Intrapersonal

CONCEPTS

Students will do the following:

1. Collect data that indicates the number of letters in their first names.
2. Find the class mean or average.
3. Calculate the absolute values or differences between individual names and the mean.
4. Use data to problem solve how many boxes should be designed on a form to ensure that there are enough boxes for most of the population in the United States to write their first names.

MATERIALS

- Copy of Activity Sheet 1 for each student
- Copy of Activity Sheet 2 for each student
- Copy of Grading Matrix for each student
- Overhead transparency of Class Chart
- Strip of one-inch squares for each student
- Scissors
- Calculators

WHAT TO DO

Discuss with students each of the questions posed on Activity Sheet 1. Give each student a strip of one-inch squares so they can write their first names in the squares (one letter in each of the squares). Place the strips of names on a table that is large enough to accommodate them so that each strip is visible. Ask students, "By looking at the strips of names, can you predict the average (or mean) number of letters in our names?" Be sure the students have an opportunity to estimate the mean.

Once students have made their estimates, explain to them that the mean can be approximated by "evening out" the strips. In other words, we can cut letters from the longer names and add these letters to the shorter names to find the mean. As this is being done, query what happens if the columns are not completely even. Ask students between which two numbers the mean will be found. Have them approximate whether the mean will be closer to one of the numbers or the other. Why do they believe this?

To calculate the actual deviation from the mean, use the data on the Class Chart. Each student writes his or her name and the number of letters in his or her first name. When this is completed, the mean for the class is computed. This number is written next to each student's data (in the column labeled $\overline{x}$). To find the absolute value of the deviation from the mean, students perform the following operation: $| x_1 - \overline{x} |$. The mean deviation can then be calculated by finding the average of these scores.

Students then form into groups of two or four and work as groups to answer the questions posed on Activity Sheet 2. This exercise requires students to analyze the data and decide how many boxes or spaces should be allocated on a form to ensure that the majority of U.S. residents would be able to write their names in the squares and not run out of boxes.

Although this procedure falls short of requiring students to find the standard deviation, they do determine a number that is a good approximation of the standard deviation, which can be used in much the same manner. Let's suppose that the deviation from the mean calculates out at 3.4; students might predict that three or four times that number might be an adequate number of spaces. Ask groups to explain their thinking and reasoning in determining the necessary number of spaces.

VARIATION

More advanced students can compute the actual standard deviation using this formula:

$$\sigma = \sqrt{\frac{1}{n-1}\sum_{i=1}^{n}(x_1 - \overline{x})^2}$$

where n = the number of pieces of data, x_1 = individual data, and $\overline{x}$ is the mean of the class.

TIE TO TECHNOLOGY

Advanced statistical analysis can be done using spreadsheet programs. These programs have extensive libraries of built-in statistical functions. Data can be entered into the spreadsheet and, using available formulas, students can statistically analyze the data.

ASSESSMENT

1. Observation and questions: "Can you explain why your group used the number of squares that it did?"

2. Journal question: "Why did we approximate the mean or average of the class when we 'evened out' the names on the table?"

3. Grading matrix

How Long Is Your First Name?

Activity Sheet 1

How many letters do you have in your first name? Is it a short or long name? When you fill out a form that asks you to put the letters of your first name in little boxes, do you run out of boxes or are there some boxes left over? How do you think the designers of these forms know how many boxes they will need so that most people will have enough room to fit in their whole name? Let's conduct two experiments to see if *statistics* can answer this question.

Directions

1. Use a strip of one-inch squares to write your first name. Pretend you are filling out our imaginary form and put one letter in each of the squares.
2. Place your name on the Class Chart underneath the names of the other members of the class.
3. Predict the average number of letters in the first names of the students in your class.
4. Pay close attention as the teacher physically "evens-out" the names by cutting letters off the longer names and adding these letters to the shorter names.

The average or mean number of letters in the first names of the students is _____

Now let's complete the information on the Class Chart so we can analyze the data we have using a statistical procedure called *deviation from the mean.*

How Long Is Your First Name?

Class Chart

Name of Student	Number of Letters in First Name (x_1)	Mean or Average of the Class ($\bar{x}$)	Difference $\|x_1 - \bar{x}\|$
Average			

How Long Is Your First Name?

Activity Sheet 2

Directions

Use the collected data and the calculated statistics to design a form with enough spaces to accommodate most of the first names of people in the United States. Work with your group to answer these questions and design the form.

1. What was the class mean? _____
2. How many more letters than the mean did the name with the most number of letters have? _____
3. How many times larger than the mean is this name? (Round to the nearest whole number.) _____
4. Problem solve the number of spaces you think the form would need to accommodate the number of letters in the first names of most of the people in the United States. Use the space below to explain your thinking.

Our Thinking:

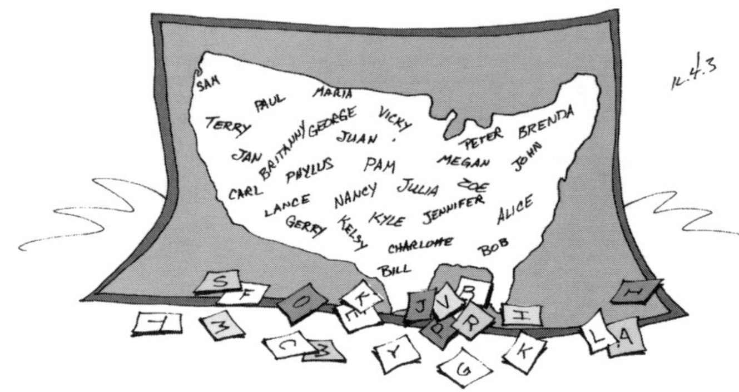

How Long Is Your First Name?

Grading Matrix

Names ____

Date____________________ Class ____________________________

Criteria	**4**	**3**	**2**	**1**
Accuracy of calculations				
Group completed the problem-solving assignment				
Quality of reasoning of number of spaces needed				
How well did group work together?				

Comments:

4 = Superlative, 3 = Competent, 2 = Limited, 1 = Inadequate

ACTIVITY 25

Shoe Length versus Height

MATH TOPICS

Data Collection and Analysis, Scatterplots, Positive Correlation (Direct Variation), Indirect Variation, Measurement, Ratios, Constants, Problem Solving

TYPES OF INTELLIGENCES

Logical/Mathematical, Bodily/Kinesthetic, Interpersonal, Intrapersonal, Verbal/Linguistic

CONCEPTS

Students will do the following:

1. Measure to the nearest $^1/_{10}$ cm.
2. Determine the horizontal and vertical units for the graph.
3. Graph the data on the coordinate plane.
4. Analyze the scatterplot to ascertain if their data shows a correlation and, if so, whether it is positive or negative.

MATERIALS

- Copy of Data Collection Sheet for each student

- Copy of Graph of Data for each student
- Copy of Grading Matrix
- Overhead transparency of Graph of Data
- Meter sticks or metric tape measures, metric rulers

WHAT TO DO

Ask students, "Do you think there is a relationship between the size of persons' shoes [i.e., shoe lengths] and their height? Why?" After there is some discussion, discuss correlation and what a graph of a positive correlation (or direct variation) might look like. Tell the students that they are going to conduct an experiment to see if there is a relationship between these two variables—shoe length and height.

Students can work in groups of four to facilitate taking measurements. Explain that each measurement should be rounded to the nearest tenth of a centimeter. When all of the students have completed their measurements and recorded them on the class Data Collection Sheet, students should copy the data on to their individual copies of the Data Collection Sheet so they can use the data to produce a graph and calculate the ratio of height to shoe length.

Have students find the average for the class. Does this number appear to be somewhat constant for the group? In other words, is it possible to predict (approximately) a person's height if one knows the length of his or her shoe? Or, if one knows the length of someone's shoe, could one predict how tall that person is? If students believe that they could make such a prediction, then the ratio they found is a mathematical constant (of sorts).

Problem solve with the students how the units on the horizontal and the vertical axes of the group should be computed. To compute, students will need to find the range of the data and the number of student scores to be recorded, then calculate units that are of equal value. It will be necessary to compress the graph between zero and the first unit on each of the axes. Allow students to calculate their own units and graph the data. Demonstrate the procedures on the Graph of Data overhead transparency.

When the graphs are completed, ask the students if the coordinates appear to be graphed in any particular direction; for example, from the upper left to the lower right or from the lower left to the upper right. Or perhaps the students see no pattern and the coordinates are randomly placed on the coordinate plane. There should appear to be a positive correlation; for example, as the height of a student increases so does the size or length of his or her shoe.

VARIATION

Students, working in groups, can design an experiment in which they believe there would be a negative correlation.

TIE TO TECHNOLOGY

This activity requires students to collect two pieces of data and, by graphing them on a scatterplot, determine if there is a positive correlation. Once they understand these concepts, students can enter their data into a spreadsheet and use the computer's graphing capabilities to create the scatterplot.

Students also can research the heights and weights of the players on their favorite basketball or football team using the Internet and graph this data to see if there is a correlation.

ASSESSMENT

1. Student observation and questioning: "Describe what an alien from another planet might look like. How would you recognize him or her as nonhuman?" or "Explain what we mean when we say, 'That person is out of proportion.'"

2. Journal question: "Discuss what is meant by a positive or negative correlation. Give an example of each type of correlation. Be sure to be as precise as possible."

3. Grading matrix

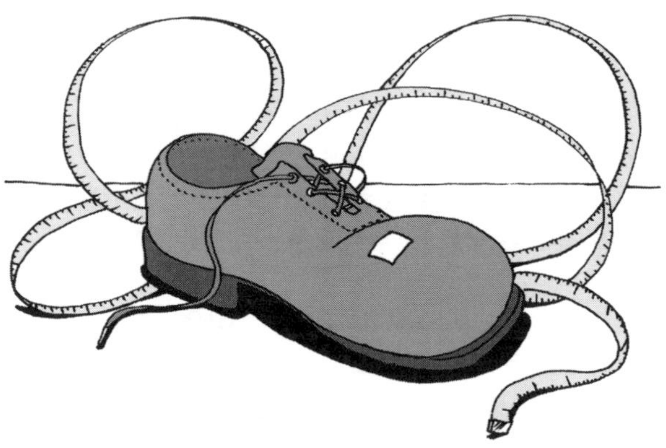

Shoe Length versus Height

Data Collection Sheet

Directions

Today we are going to conduct an experiment to see if there is a relationship between the size of our shoes and our height. Measure your foot (wearing shoes) and your height to the nearest $^{1}/_{10}$ of a centimeter. Record your data on the table below.

Name of Student	Height (cm)	Shoe Size (cm)	Ratio Height/Shoe Length

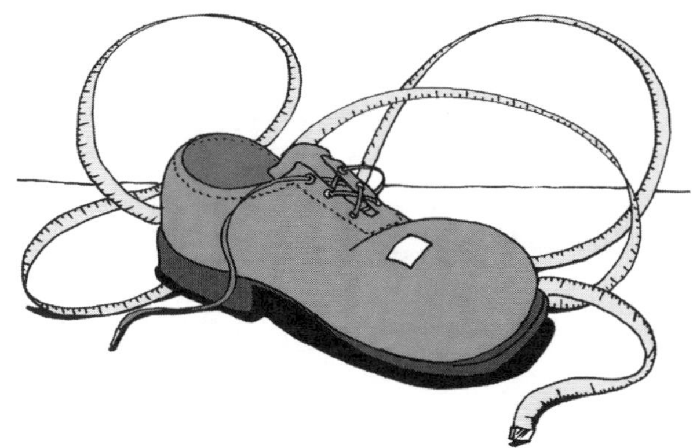

Shoe Length versus Height

Graph of Data

Class Data

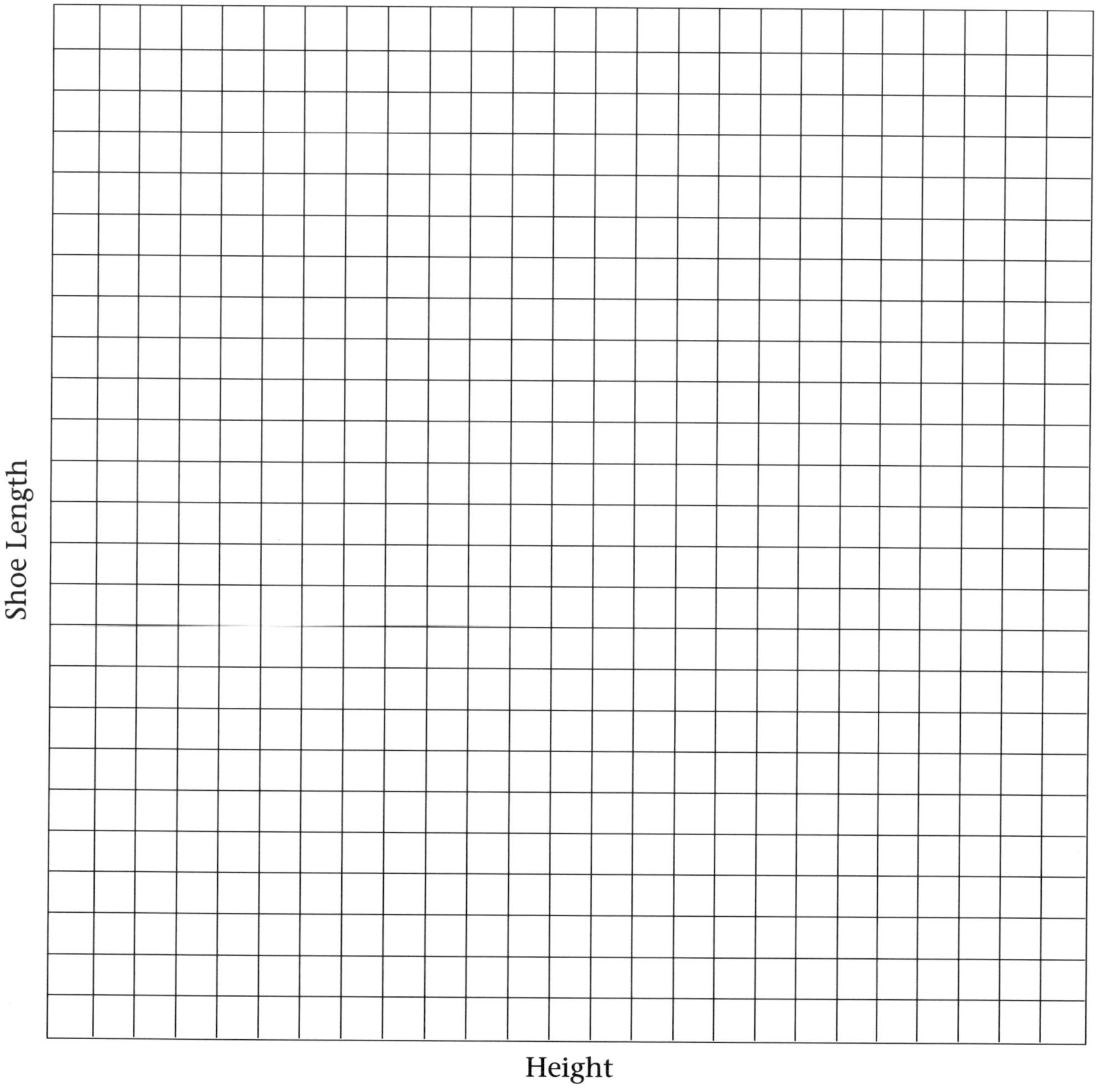

Shoe Length

Height

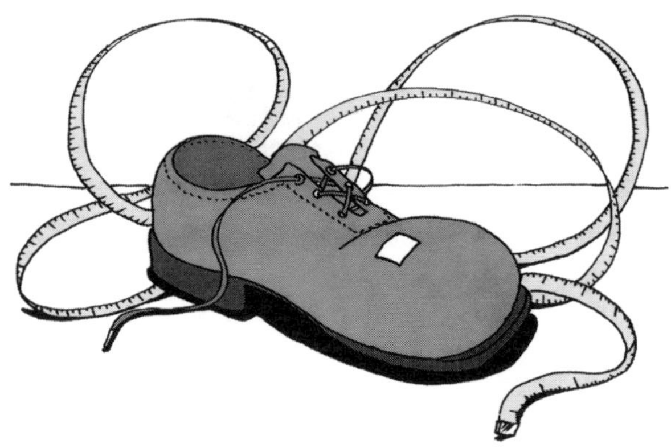

Shoe Length versus Height

Grading Matrix

Names ______________________________

Date__________________ Class ______________________________

Criteria	**4**	**3**	**2**	**1**
Accuracy of measurement				
Correctness of graphing units				
Overall quality of graph				
How well did group work together?				

Comments:

4 = Superlative, 3 = Competent, 2 = Limited, 1 = Inadequate

ACTIVITY 26

Millions of Marshmallows

MATH TOPICS

Data Collection and Analysis, Sampling, Estimation, Ratio and Proportion, Percentage of Difference

TYPES OF INTELLIGENCES

Logical/Mathematical, Interpersonal, Intrapersonal, Verbal/ Linguistic

CONCEPTS

Students will do the following:

1. Estimate the number of miniature marshmallows in a bag.
2. Work in groups of four to conduct sampling experiments to more accurately predict the number of marshmallows in their bag.
3. Find the average number of marshmallows in each of the group's bags.
4. Use the data collected to predict the range of marshmallows they might expect to find in an average-sized bag of marshmallows.

MATERIALS

- Copy of Activity Sheet for each group
- Overhead transparencies of Data Collection Table (see Activity Sheet) and Class Results
- Bag of white miniature marshmallows for each group
- Bag of colored miniature marshmallows for the class to share
- Large paper bag to mix marshmallows
- Cup for each group to scoop out samples of marshmallows
- Calculators

WHAT TO DO

Have students form into groups of four to conduct this experiment. Then, hold up a bag of miniature marshmallows and ask students to decide, as a group, how many marshmallows they think are in the bag. Each group should record its estimate on the Class Results transparency and in the space provided on their own Data Collection Tables (part of Activity Sheet).

Students, working in groups, replace 50 of the white marshmallows in their bag with 50 of the colored marshmallows. These marshmallows are now considered "tagged" and can be released "into the wild." Students should carefully mix the "tagged" marshmallows with "the rest of the population" of marshmallows. All should be dumped into a large paper bag so they can be mixed well after each draw.

Using a cup, students take turns scooping out a sample of the marshmallow "population." They count the colored marshmallows and then the total number of marshmallows in their cup (white and colored), recording their data on the data collection table. Each group conducts ten trials and records the data for each trial. Finally, an average of each of the trials is calculated and recorded.

Using the ratio:

$$\frac{\text{average number of marshmallows}}{\text{average total number}} = \frac{50}{x}$$

By using cross-products, students can calculate the number of marshmallows in their bag. When all groups have finished, they should record their totals on the Class Results transparency and find an average for the class.

VARIATION

Students can research how sampling is used by scientists in ways other than population studies. Or, students can explore the controversy that developed when the U.S. government census takers wanted to use this technique to obtain the number of difficult-to-count populations in the 2000 census.

ASSESSMENT

1. Observation and questioning: "Can you explain whether this technique will help you predict the number of marshmallows in the bag? Why or why not?"
2. Journal question: "Describe the statistical technique of sampling that was used in the marshmallow experiment. Do you think it is an accurate way to make predictions? Why or why not? Be sure to justify your opinion."
3. Grading matrix

ON THE INTERNET

Students can learn more about statistics and probability through sites available on the Internet. The Math Forum's topics page (see appendix for URL) provides links to other Internet sites that offer activities, weekly challenges, and related information.

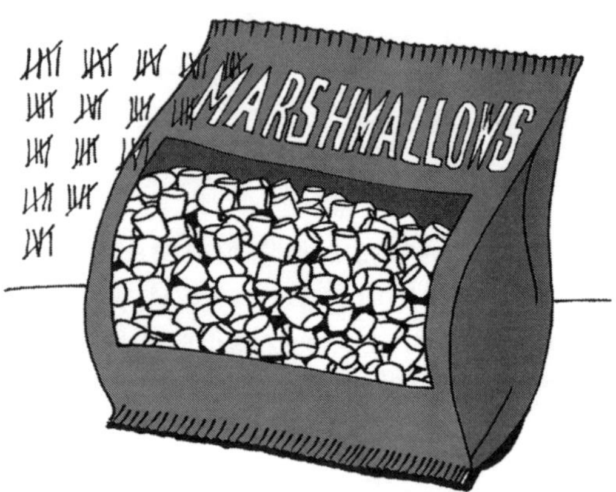

Millions of Marshmallows

Activity Sheet

Just how many marshmallows are there in a big bag? Would you like to be given the job of counting them? Probably not!

Sometimes we want to know how many of something there is but the number either is so large that we don't want to count them or it would be impossible to count even if we wanted to. Can you think of some instances where we might be faced with either of these two alternatives?

We will be conducting an experiment that will help us estimate the number of marshmallows in this bag quite accurately. The process is called capture-recapture and it is known as experimental sampling. Here's what we will do:

1. First predict how many marshmallows are in the bag. Write your estimate here __________
2. Next, take 50 white marshmallows out of the bag and set them aside.
3. Place 50 colored marshmallows in the bag.
4. Mix the colored and remaining white marshmallows thoroughly.
5. Scoop out a sampling of marshmallows. Count the colored marshmallows and the total of all the marshmallows in the scoop and record both numbers on the Data Collection Table (next page).
6. Put all the marshmallows back and shake the bag well.
7. Do this 9 more times (a total of 10 trials).

Directions

Use the averages on this table to estimate the number of marshmallows in your bag. Use the proportion shown (and your data) to calculate the estimate. Then count the marshmallows in the bag to see how close your estimate was.

Data Collection Table

Trials	Number of Colored Marshmallows	Total Number of Marshmallows
1		
2		
3		
4		
5		
6		
7		
8		
9		
10		
Average		

$$\frac{\textbf{average number of marshmallows}}{\textbf{average total number}} = \frac{50}{x}$$

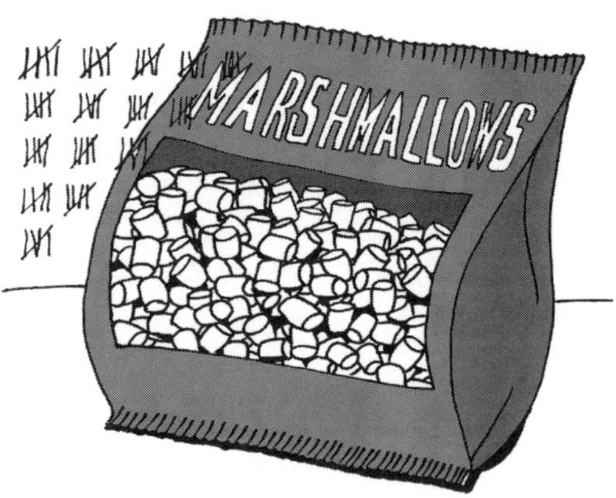

Millions of Marshmallows

Class Results

Group	Predicted # of Marshmallows	Actual # of Marshmallows	% of Difference
Class Average			

Based on the data we collected in our experiments, we believe that there are between __________ and __________ marshmallows in an average-sized bag of miniature marshmallows.

Millions of Marshmallows

Grading Matrix

Names ____

Date____________________ Class ______________________________

Criteria	4	3	2	1
Accuracy of data collection				
Accuracy of computation				
How well did group work together?				
Comments:				

4 = Superlative, 3 = Competent, 2 = Limited, 1 = Inadequate

ACTIVITY 27

The Case of the Disappearing Telephone Numbers

MATH TOPICS

Combinations, Permutations, Data Collection and Analysis, Graphing

TYPES OF INTELLIGENCES

Logical/Mathematical, Verbal/Linguistic, Interpersonal, Intrapersonal

CONCEPTS

Students will do the following:

1. Calculate the number of available telephone numbers using concepts of permutations.
2. Survey their classmates to determine the number of telephone numbers being used by individuals.
3. Calculate the average number of telephone numbers being used per household.
4. Using an analysis of their data, predict the availability of new telephone numbers in their community.

MATERIALS

- Copy of Activity Sheet for each student
- Copy of Telephone Number Survey for each student

- Copy of Survey Analysis for each student
- Copy of Grading Matrix for each student
- Calculators

WHAT TO DO

Discuss the problem communities across the United States are facing with the advent of ten- and eleven-digit telephone numbers. There is a great deal of controversy and questions about the need to change area codes, ten- and eleven-digit overlays, and possible eight-digit local telephone numbers. Ask students if they believe the telephone companies when they say we are running out of numbers! There is more here than *meets the eye!* It has been reported that up until six years ago, the average household had one telephone number. Today the number is between two and three and is rapidly approaching three. Because new numbers are assigned to fax machines, pagers, and cellular phones, many communities now have to dial ten or eleven digits to dial a number across town!

How many telephone numbers can be distributed? For the standard seven-digit number:

2–9, 0–9, 0–9, 0–9, 0–9, 0–9, 0–9 or
$8 \times 10 \times 10 \times 10 \times 10 \times 10 \times 10$

8,000,000 telephone numbers are available.

The question to ask is, "If there are so many telephone numbers, why are we running out of numbers?" To answer this question, students survey 25 people and record how many telephone numbers each of these persons has in his or her home and how these numbers are being used (e.g., fax, business phone, personal phone). By finding the average number of phones and applying that number to the population of their city, students can predict the number of telephone numbers in use in their town. Population information can be obtained on the Internet, in published census reports, or by contacting local officials.

To expedite authentic assessment, a written assignment is required of each student. They are asked to evaluate their survey data and indicate how this data affects the availability of new telephone numbers in their city.

VARIATION

Many states currently have lottery drawings. Students can explore the probability of winning the lottery using either combinations or permutations. Most Lotto drawings are based on combinations; daily lottery numbers must be in the same order they are called, and so these are based upon permutations.

TIE TO TECHNOLOGY

Students survey 25 people and record the telephone equipment they use in their households. After organizing the data, it can be entered into a spreadsheet. Percentages can be calculated and appropriate graphs can be created. The project can be described using a word processing program and the written description, tables, and graphs can be imported into a PowerPoint report for an oral presentation.

ASSESSMENT

1 Quality of data collection as shown on Activity Sheet
2. Written analysis of data
3. Journal question: "When the telephone company requires ten-digit dialing for every call, how many telephone numbers does that add to the total number? Remember that the first digit cannot be a zero or a one. Explain how you got your answer."
4. Grading matrix

ON THE INTERNET

For information on area codes, how areas have been split, where area codes are located, and related information, the website LincMad (see appendix for URL) offers a vast array of information. This site provides some discussion about possible eight-digit local numbers to eliminate the problem of a lack of numbers.

The Case of the Disappearing Telephone Numbers

Activity Sheet

All over the United States, people are receiving the news that we're running out of telephone numbers! How can that be? What's the problem here? Well, the telephone companies are saying that there are a limited number of telephone numbers and with cellular phones, pagers, and fax machines, these numbers are being used up!

Let's examine how many numbers actually are available for the standard seven-digit telephone number. First, there is one very important rule: One and zero cannot be used as the first digit—because if we dial a zero, the operator will pick up the call, and if we dial a one, the call will be rerouted to a long distance phone line!

Use the blank spaces below to help calculate the total number of possible telephone numbers:

_____ _____ _____ - _____ _____ _____ _____

There are ten digits available to use in telephone numbers: 0, 1, 2, 3, 4, 5, 6, 7, 8, and 9.

Since we cannot use a zero or a one as the first digit, how many digits are available for the first number? _____

How many digits are available for the second number? _____

The third? _____ the fourth? _____ the fifth? _____ the sixth? _____ the seventh? _____

How many different telephone numbers are available if we have a seven-digit dialing number? _____

To understand why we are running out of telephone numbers, we need to have some additional information:

- On average, how many telephone numbers do people in your town or city have?
- How many people are there in your town or city? You may use the Internet or other census information to obtain this information.

To help gather the necessary information, why not take a survey? Then use this data to estimate the number of telephone numbers the average person in your town has. (We will assume that the people you surveyed are a representative sample of the town's population.) Use the following data collection table to record your findings.

The Case of the Disappearing Telephone Numbers

Telephone Number Survey

Questions

1. How many different telephone numbers do you have in your household?
2. What equipment are these numbers connected to; for example, pagers, fax machine, cellular phones, etc.?

Directions

Survey 25 people and record their information on the table below. Then organize the data and write an analysis of the average number of telephone numbers being used in these households.

Name of Person	Number of Telephone Numbers	What Equipment Is Being Used

The Case of the Disappearing Telephone Numbers

Survey Analysis

To determine if your town or city will be running out of telephone numbers because of increased usage, you have researched the number of people in your town and have analyzed the results of your survey. Write a "newspaper article" describing the results of your survey and your impressions regarding the availability of telephone numbers.

The Case of the Disappearing Telephone Numbers

Grading Matrix

Names ______________________________

Date_______________ Class ______________________________

Criteria	4	3	2	1
Accuracy of calculations				
Quality of collection and organization of data				
Overall quality of analysis				
Quality of "newspaper article"				

Comments:

4 = Superlative, 3 = Competent, 2 = Limited, 1 = Inadequate

Appendix A

A Collection of Math Medleys

÷The Divisibility Ditty÷

Sung to "The Itsy Bitsy Spider"

When looking for the factors
Of any number please,
Check out the rules,
Then knowledge comes with ease.
Seek out only evens
Odds don't rate a look.
When dividing by two, my friend,
We'll go right by the book.

To work with three is fun,
It's rule is not too bad.
Sum up the digits
And see what sum you had.
If three goes into it,
It surely does you say,
Then the number that you started with
By three it was okay.

Now four has other rules to watch
Don't look at odds, no way.
Check out the evens
And one more thing we say.
The last two digits,
Divisible by four.
That's the rule complete and true
We cannot tell you more.

Now what to say about the fives?
A fine and noble friend.
Fingers on one hand
And nickels we do spend.
Look out for fives and zeros,
These numbers it must end.
Divisibility by five
Is certain my good friend.

Oh, six is odd and even too,
It takes two rules to solve.
Rules for two and three
This problem will resolve.
First it must be even,
And then divide by three,
Divide by six is possible,
It's mathematically!

Divide by eight is just like four,
The rule is much the same.
Check just the evens,
Dividing is the game.
Check the last three digits
Divisible by eight.
Another rule is in the bag,
Can we keep all this straight?

To solve for nine just look at three,
Their sameness they can't hide.
Just find the sum
By nine they must divide.
There's no remainder.
These numbers do abound,
Fun with tunes and meter,
The rule for nines we've found.

The last to find—the number ten,
It's rule is short and sweet.
Zeroes alone do
End these numbers neat.
We've learned all the rules of
Di-visi-bil-ity.
From 2 to 4 to 6 to 10
Start over now, oh gee!

O^3—Order of Operations Song

Sung to "Frère Jacques"

Do a problem, in what order?
What to do? There are rules.
First we do parentheses, they're the first to look at.
We're no fools, use the rules.

Next exponents, roots, and powers,
Once that's done, what comes next?
One step for each lesson, now we're on a mission.
We're no fools, use the rules.

Mul-ti-pli-ca-tion or di-vi-sion,
They come next, left to right,
Adding and subtracting, ends our little problem.
"O" to the third, we've been heard!

P	**E**	**D**	**M**	**A**	**S**
l	v	o	a	n	m
e	e		t	d	i
a	r		h		l
s	y				e
e	o				
	n				
	e				

The Congruence Chorus

Sung to "Three Blind Mice"

Con-gru-ence, con-gru-ence,
Geometric twins, geometric twins,
Let's look at triangles, there are three,
All sides the same length—they're sure to be,
Known as side-side-side—it's just one of the three,
That's congruence.

Con-gru-ence, con-gru-ence,
Geometric twins, geometric twins,
Two sides and their angle—a second one,
The angle is caught in the middle, such fun,
Known as side-angle-side—the second of three,
That's congruence.

Con-gru-ence, con-gru-ence,
Geometric twins, geometric twins,
Two angles one side is the third, you see,
Sides caught in the middle, oh my oh me,
Known as angle-side-angle, the last of the three,
That's con-gru-ence.

SSS—SAS—ASA

Ode to Pythagoras

Sung to "Hokey Pokey"

You take your first leg "a,"
And then your next leg "b,"
Take the sum of their squares,
Are you following me?

To use this famous theorem,
For Pythagoras let's shout,
That's what it's all about.

We're not quite finished yet,
There's a third side to see,
'Cause this tri-angle's right,
Are you following me?

To use this famous theorem,
For Pythagoras let's shout,
That's what it's all about.

Now add the square of leg "a,"
To the square of leg "b,"
You get the square of side "c,"
Are you following me?

To use this famous theorem,
For Pythagoras let's shout,
That's what it's all about.
$a^2 + b^2 = c^2$

Rounds of LCM & GCF

Sung to "Row, Row, Row Your Boat"

LCM my friend
Is easy to define,
For any two numbers that you may choose,
Mul-ti-ples we'll find.

Take the number four,
And check it out with six,
The first common multiple we find is twelve,
The LCM—no tricks.

What if the numbers are,
Three and five this time?
For their LCM we just mul-ti-ply,
They're rel-a-tive-ly prime.

Now you pick your own,
Two numbers be so kind,
Work with a friend to problem solve,
The LCM to find.

GCF is just,
A version of the first,
Instead of searching for multiples,
Factors are our thirst.

Take the number four
And check it out with six,
Their only factor in common is two,
Their GCF—no tricks.

If the numbers are
Three and five this time,
Their GCF is one because,
They're relatively prime.

What of eight and twelve?
Our choice is two or four.
The greatest factor in common we seek,
It's FOUR—there are no more.

LCM is done,
So is GCF,
To learn them both took lots of time,
But let's become adept!

Appendix B

Surfing the Web

When surfing the web, there are a few important points to keep in mind:

- Find a search engine that allows you the greatest margin for error. Some of the larger (and more user-friendly) search engines are Yahoo, Metacrawler, Lycos, Snap, Webcrawler, and Excite. Experiment—it's fun! Following are their URLs.

  — Yahoo: http://www.yahoo.com

  — Metacrawler: http://search.go2net.com/crawler or http://www.metacrawler.com

  — Lycos: http://www.lycos.com

  — Snap: http://www.snap.com

  — Webcrawler: http://www.webcrawler.com

  — Excite: http://www.excite.com

- Be as specific as possible when using keywords in your search. If you use a keyword such as *mathematics,* you will get a list of so many sites that you probably won't have the time to find what you are looking for. Narrow the search as much as possible—if you want to find information on prime numbers, you might go to a mathematics education site and then type in the search term *prime numbers.*

- Focus on sites that you know will contain accurate information. These might be educational sites (*edu* will appear in the name) or government sites (*gov* will appear in the name). There is a great deal of misinformation out there and sources need to be verified.

The following sites relate to the activities and projects presented in this book. They will provide the means for students to expand and explore the Internet, a truly remarkable information highway.

GENERAL

A Great Math Site

A website jam packed with a multitude of mathematics topics is The Math Forum at <http://forum.swarthmore.edu/>. It contains activities to use in the classroom, research and information about math history, and product references. New topics and material are regularly added, so one should check it frequently. Its topic page is <http://forum.swathmore.edu/math.topics.html>.

Chapter One

Activity 2: All About the Moon
The website The Earth and Moon Viewer shows students the position of Earth and the moon relative to their own latitude, longitude, and altitude. Its URL is <http://www.fourmilab.ch/earthview/vplanet.html>.

A much more comprehensive site that explains in great detail the moon and planets and space explorations to the moon is <http://seds.lpl.arizona.edu/nineplanets/nineplanets/luna.html>.

Activity 3: Cooking and Mathematics: Converting Recipes
Check out all of the wonderful food sites on the Internet. Typing the keyword *recipes* into the search engine Webcrawler produced over 11,000 websites to explore. Some of the more interesting ones are cited below:

A site that contains recipes from around the world (including ancient Roman and medieval recipes is <http://www.mit.edu:8001/people/wchuang/cooking/cooking.html>. Another site, <http://www.mit.edu:8001/people/wchuang/cooking/recipes/Conversion.txt>, converts units of measure from metric to standard for hundreds of items.

The cable Food Network's website is <http://www.foodtv.com/fn/recipes/>. Students and teachers can email questions they might have about mathematics and cooking.

A site with great recipes is <gopher://ftp.std.com/11/obi/book/HM.recipes/TheRecipes>. It contains hundreds of recipes, listed alphabetically from almond berry gateau to zabaglione. It is a no-frills site but is loaded with recipes!

And a site with recipes for every imaginable cookie is CookieRecipe.com at <http://www.cookierecipe.com>.

Activity 5: The Pattern Tells It All
The Math Forum: <http://forum.swarthmore.edu/>

Activity 6: Eratosthenes and the 500 Chart
The Prime Page: < http://www.utm.edu/research/primes>

The University of Utah page titled Eratosthenes of Cyrene: <http://www.math.utah.edu/~alfeld/Eratosthenes.html>

Activity 7: Math and Music
There are many very exciting websites connected to the learning and teaching of mathematics. One that relates to music and math is <http://tqjunior.advanced.org/4116/Music/music.htm>. It shows the relationship between Fibonnaci numbers and music and discusses recent research that attributes higher math scores with the study of piano and much more.

Chapter Two

Activity 9: The Irrational Spiral
An interesting website that has an animated version of the proof of the Pythagorean Theorem is <http://www.nadn.navy.mil/MathDept/mdm/pyth.html>.

Activity 10: Vet Math
American Kennel Club: <http://www.akc.org/>

Activity 11: Seeing to the Horizon
A Digital Archive of American Architecture: <http://www.bc.edu/bc_org/avp/cas/fnart/fa267/20_sky.html>

Activity 12: Falling Objects
World of Coasters: <http://www.rollercoaster.com/>

Activity 13: The 18-Hour Clock
Symmetry and Pattern: The Art of the Oriental: <http://forum.swarthmore.edu/geometry/rugs/>

Chapter Three

Activity 17: Math and Music: Frequencies
A website devoted to explaining the relationship between music and math and science is <http://www.math.niu.edu/~rusin/papers/uses-math/music/>. The site explains the nature of sound and frequencies found in an octave, and provides suggestions for high school projects.

MidiWeb Information Guide, a website that has been set up to help musicians learn the technology of music, can be found at <http://www.midiweb.com/index2.html>. It contains updated news, technical information, links to other sites, and tips for programming.

Activity 19: The Valley of Mars
Moon and Planets: There are many Internet sites on the planets, and in addition to their useful math facts, students will enjoy the beautiful astronomical photography that has been added to them.

An interactive website designed by the Adler Planetarium in Chicago is <http://www.adlerplanetarium.org/>. This site is updated regularly, with new information and programs frequently added. It is a useful site to bookmark for future updates.

Two of the most beautiful astronomical sites on the Internet are Welcome to the Planets at <http://pds.jpl.nasa.gov/planets/> and NASA's Planetary Photojournal at <http://photojournal.jpl.nasa.gov/>. Both have breathtaking pictures of the planets and show their relative sizes in its opening graphic. NASA's site allows viewers to select the types of pictures they want to view, whether through the Hubble telescope or as pictures taken on space voyages. Amazing! At both sites, students can access all of the planets for research purposes.

Students for the Exploration and Development of Space (SEDS) is a wonderful website just loaded with information. Students can take a tour of the nine planets or keep track of the space shuttle schedule. Its URL is <http://www.seds.org/>.

The International Association of Astronomical Studies is a group of middle school and high school students in the Denver, Colorado area who maintain a website to share ideas with other students interested in pursuing astronomy and space. This site can be found at <http://www.iaas.org/>.

The Mt. Wilson Observatory website (<http://www.mtwilson.edu/index.html>) contains the latest news as well as some interesting ideas to bring astronomy into the classroom.

Activity 20: Limitless Lung Power
Interesting mathematics activities can be found at <http://www.nlhep.org/testing2.htm>. This site is sponsored by the National Lung Health Education Organization and a variety of medical groups. It describes some interesting tests students can perform to find their lung capacities.

Activity 21: Tessellations: Transformations a la Escher
One of the more beautiful websites dedicated to the teaching and learning of tessellations is Totally Tessellated (<http://library.advanced.org/16661/>). It contains areas that discuss the history and essentials of tessellation, as well as M. C. Escher and his art. It is a wonderfully informative site.

Another interesting site is <http://forum.swathmore.edu/sum95/suzanne/tess.html>. A student, using HyperCard, describes how to create tessellations using a computer.

Chapter Four

Activity 22: Predicting Colors in a Bag of M&Ms
Mars, Incorporated: <http://www.m-ms.com/>

Activity 23: Phone Home
See sites listed under Chapter Three, Activity 19.

Activity 24: How Long Is Your First Name?
The official site of the Census Bureau is <http://www.census.gov/>. This site contains a wealth of information about population growth and trends. Another site, <http://www.fedstats. gov/map.html>, allows students to click on a map of the United States and find information about their own state.

Pop Clock: Provides information on the U.S. Census Bureau and statistical measures. <http://www.crpc.rice.edu/CRPC/GT/sboone/Lessons/Titles/popclock.html>

Activity 25: Shoe Length vs. Height
Students can learn the heights and weights of all of the players on their favorite teams. <http://www.nba.com/team_roster.html> is the official site for the National Basketball Association. <http://www.nfl.com/players/> is the official site for the National Football League.

Activity 27: The Case of the Disappearing Telephone Numbers
LincMad: http://www.lincmad.com/

Bibliography

Anderson, S., Ball, S., Murphy, R. and Associates. 1975. *Encyclopedia of Educational Evaluation.* San Francisco: Jossey-Bass.

Armstrong, T. 1993. *Seven Kinds of Smart.* New York: Penguin.

Armstrong, T. 1994. *Multiple Intelligences in the Classroom.* Alexandria, VA: Association for Supervision and Curriculum Development.

Bolster, C. 1973, April. Activities: Tessellations. *Mathematics Teacher* 66: 339–342. Reston, VA: National Council of Teachers of Mathematics.

Bool, F. H., Kist, J. R., Locher, J. L., Wierda, F. 1992. *M. C. Escher.* New York: Abrams.

Chittenden, E. 1991. Authentic Assessment, Evaluation, and Documentation of Student Performance. In *Expanding Student Assessment,* edited by V. Perrone. Alexandria, VA: Association for Supervision and Curriculum Development.

How Escher Created. 1986, December/January. *Art and Man,* 2–3.

Poet of the Impossible. 1986, December/January. *Art and Man,* 2–3.

Gardner, H. 1983. *Frames of Mind: The Theory of Multiple Intelligences.* New York: Basic Books.

Gardner, H. 1995, November. Reflections on Multiple Intelligences. *Phi Delta Kappan.*

Lawson, D. R. 1990. The Problem, The Issues That Speak to Change. In *Algebra for Everyone,* edited by E. L. Edwards Jr. Reston, VA: National Council of Teachers of Mathematics.

Locher, J. L. 1971. *The World of M. C. Escher.* New York: Abrams.

Martin, H. 1996. *Multiple Intelligences in the Mathematics Classroom.* Palatine, IL: IRI/SkyLight.

Mathematical Sciences Education Board, National Research Council. 1989. *Everybody Counts.* Washington, DC: National Academy Press.

National Council of Teachers of Mathematics. 1989. *Curriculum and Evaluation Standards for School Mathematics.* Reston, VA: author.

National Council of Teachers of Mathematics. 1995. *Assessment Standards for School Mathematics.* Reston, VA: author.

National Council of Teachers of Mathematics. 1998. *Principles and Standards for School Mathematics: Discussion Draft, Standards 2000.* Reston, VA: author.

National Research Council. 1989. *Everybody Counts.* Washington, DC: National Academy Press.

Pappas, T. 1995. *The Music of Reason.* San Carols, CA: Wide World Publishing.

Stenmark, S. S. 1990. *Assessment Alternatives in Mathematics.* Berkeley, CA: EQUALS, University of California.

Tartre, G. C. 1990. Spatial Skills, Gender, and Mathematics. In *Mathematics and Gender,* edited by E. Fennema and G. Leder. New York: Teachers College Press.

WTB Magazine, http://www.worldstallest.com/96/fansky.html, August 2, 1999.

Index

There are
one-story intellects,
two-story intellects, and
three-story intellects with skylights.

All fact collectors, who have no aim beyond their facts, are
one-story minds.
Two-story minds
compare, reason, generalize,
using the labors of the fact collectors
as well as their own.
Three-story minds
idealize, imagine, predict—their best illumination comes from above,
through the skylight.

—Oliver Wendell Holmes